The unconscious at work

Working in the human services has always been stressful, and the current massive changes in the organization of these services in Britain and elsewhere have added to the stresses inherent in the work. *The Unconscious at Work* explores the difficulties experienced by managers and staff in a wide range of care settings, and investigates the unconscious processes that add greatly to the stress inevitable in caring work.

The authors, all past or present members of the Tavistock Clinic Consulting to Institutions Workshop, make use of ideas from psychoanalysis, open systems theory, Bion's work with groups, and group relations training. Drawing on their experience of consulting to individuals, teams and organizations, they distil what they have learned about institutional processes, and present this in an accessible and practical way. Each concept is introduced in as non-technical a way as possible, with examples from practice to make them recognizable and useful to the reader. Individual chapters develop themes relating to work with a particular client group or in a particular setting – including hospitals, schools, day centres, residential units, therapeutic communities and community care agencies. They also explore aspects of work organization – for example, the supervisory relationship, facing cuts and closure, and intergroup collaboration. In each situation the authors describe both the difficulties for which consultation was requested and their own feelings and thoughts while consulting to these institutions.

The Unconscious at Work is designed for people managing and working in the human services, and it offers readers new ways of looking at their own experiences of stress at work. It will greatly increase their understanding of the processes which can undermine effectiveness and morale, and will also be of value to consultants, trainers, and students of organizational behaviour.

Anton Obholzer is Chief Executive of the Tavistock and Portman Clinics, London, Consultant Psychiatrist at the Tavistock Clinic and a practising psychoanalyst. **Vega Zagier Roberts** is an organization consultant, a psychotherapist in private practice and a Visiting Tutor at the Tavistock Clinic.

The unconscious at work

Individual and organizational stress
in the human services

By the Members of the Tavistock Clinic
Consulting to Institutions Workshop

Edited by Anton Obholzer
and Vega Zagier Roberts

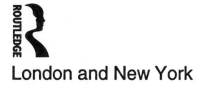

London and New York

First published 1994
by Routledge
11 New Fetter Lane, London EC4P 4EE

Simultaneously published in the USA and Canada
by Routledge
29 West 35th Street, New York, NY 10001

Reprinted 1995, 1996

Routledge is an International Thomson Publishing company

Typeset in Times by LaserScript, Mitcham, Surrey
Printed and bound in Great Britain by
Mackays of Chatham PLC, Chatham, Kent

British Library Cataloguing in Publication Data
A catalogue record for this book is available from the British Library

Library of Congress Cataloguing in Publication Data
A catalogue record for this book is available from the Library of Congress

ISBN 0-415-10205-7 (hbk)
ISBN 0-415-10206-5 (pbk)

To Eric Miller
With gratitude and affection

Contents

List of figures

About the authors

THE EDITORS

Anton Obholzer, FRCPsych., has been Chairman of the Tavistock Clinic Consulting to Institutions Workshop since its inception in 1980. Following medical training and work in general practice, he qualified in psychiatry and trained as a psychoanalyst. He is a consultant psychiatrist in the National Health Service and was Chairman of the Tavistock Clinic from 1985 to 1993. He is currently Chief Executive of the Tavistock and Portman Clinics, Associate Director of the Tavistock Institute of Human Relations Group Relations Training Programme, and has been a member of staff and Director of the Leicester Group Relations Training Programme on many occasions. He is particularly interested in consulting to institutions from a psychoanalytic group relations perspective, and has done so for a variety of public and private sector organizations, both in the United Kingdom and internationally.

Vega Zagier Roberts, MD, is Programme Organizer of the Tavistock Clinic Consulting to Institutions Workshop. She is a member of the management group of OPUS Consultancy Services, consulting to public and voluntary sector organizations, teams and individual senior managers. She is also a tutor on the Tavistock Clinic course 'Consultation, Management and Organizational Process', a psychoanalytic psychotherapist in private practice, and a member of the UK Forum for Occupational Health. She has worked as a staff member on group relations conferences, and was formerly a Fellow of the Tavistock Institute of Human Relations Action Research Training Programme.

THE AUTHORS

Wendy Bolton, B.Sc., Dip. Clin. Psych., TQAP, is a clinical psychologist and psychotherapist and Co-ordinator of Psychotherapy Services at the Penge Mental Health Centre, Ravensbourne NHS Trust. She is a member of the Society of Psycho-analytical Psychotherapists and also an independent consultant working in the selection, training and development of managers, and the management of change. She is a staff member on group relations conferences.

Francesca Cardona, BA, is an independent consultant working with public, private and voluntary sector organizations. She is a member of the management group of OPUS Consultancy Services, and associate of the Studio APS, Milan. She is also a visiting tutor at the School of Psychotherapy and Counselling, Regent's College, and a staff member of group relations conferences in the United Kingdom and Italy.

Christopher Clulow, BA, GCSW, who is Director of the Tavistock Marital Studies Institute, is a marital psychotherapist, teacher and researcher, as well as a staff member on group relations conferences. He previously worked as a probation officer, and has since offered teaching and consultancy to the probation service.

Nancy Cohn, BA, Dip. AT, MACP, is Principal Child Psychotherapist with Thameside Community Healthcare NHS Trust. She is also a tutor for the 'Diploma in Adult Counselling' course at the University of London's Department of Extra-Mural Studies. She is a member of Tavistock Clinic Visiting Teaching Staff for the Diploma/MA course in 'Psychoanalytic Observational Studies', as well as a staff member on group relations conferences.

Anna Dartington, BA, Dip. Soc. Admin., is a psychoanalytic psychotherapist and Senior Lecturer in Social Work in the Adolescent Department of the Tavistock Clinic, and a lecturer in Adult Counselling at Birkbeck College, University of London. She is a staff member on Tavistock international group relations conferences and is involved in organizational and management consultancy to groups and individuals.

William Halton, BA, is Vice-Chair of the Consulting to Institutions Workshop. He is a psychoanalytic psychotherapist and Visiting Lecturer at the Tavistock Clinic and the British Association of Psychotherapists' Child Psychotherapy Training Programme. He consults to counselling, careers and medical advice agencies, and to educational institutions.

Rob Leiper, B.Sc., MA, Ph.D., Dip. Psych., is a consultant clinical psychologist and psychoanalytic psychotherapist. He is Deputy Director of the South East Thames 'Regional Clinical Psychology Training' course and specialist in psychotherapy for Greenwich Healthcare Trust, Department of Psychology. He was formerly Projects Director for Quality Assurance at Research and Development for Psychiatry.

Chris Mawson, B.Sc., C. Psychol., is a clinical psychologist and psychoanalyst in private practice. Formerly Senior Clinical Psychologist in the Department of Child Psychiatry, Paddington Green, and the Adolescent Department, Tavistock Clinic, he has a special interest in consulting to institutions caring for children.

James Mosse, BA, is an institutional consultant and psychoanalytic psycho-therapist and a member of OPUS Consultancy Services. He consults to staff in public and voluntary sector organizations and has worked extensively as a staff member on group relations conferences. He also has ten years' experience as a manager in industry and in the voluntary sector.

Deirdre Moylan, BA, MA, Dip. Psych., TQAP, is Consultant Clinical Psycho-logist at the Adolescent Department of the Tavistock Clinic. She is also a tutor on the Tavistock Clinic course 'Consultation, Management and Organizational Process'. She consults to teams and working groups and is a staff member of group relations conferences.

Peter Speck, B.Sc., MA, Dip. Pastoral Studies, is Chaplain of the Royal Free Hospital, Honorary Senior Lecturer at the Royal Free Hospital School of Medicine, and Visiting Tutor at the Tavistock Clinic, on the course 'Consultation, Management and Organizational Process'. He has undertaken training and consultancy work with groups and individuals on issues relating to bereavement and terminal care and has worked as a staff member on group relations conferences.

Jon Stokes, MA, C. Psychol., is a psychoanalytic psychotherapist and Director of the Tavistock Clinic Consultancy Service. He was formerly Chairman of the Adult Department at the Tavistock Clinic. He also works as an independent consultant in private practice. His particular interests are role consultation to senior managers in both public and private sectors and the problems of professional partnership organizations. He has directed the 'Pride and Prejudice in Inter-Professional Work' workshops and has directed and worked on many Tavistock group relations conferences both in the United Kingdom and internationally.

Foreword

Writing this foreword offered me an opportunity to remember my own experiences in the Consulting to Institutions Workshop, which has served as a kind of incubator for the ideas developed in this volume. That was over ten years ago, and although only one of the authors represented in this volume was in the workshop at that time, the same spirit of integrity, forthrightness and compassion that touched me so then has clearly remained in force since, if the following chapters can be taken as an indication.

A central theme of these chapters concerns the need for human service professionals to confront the powerful and primitive emotional states that underlie helping relationships (especially with people in dire need), and consider how the staff members of these organizations can function effectively without becoming chaotic or withdrawn. Another is how the organizational arrangements themselves – the structures, cultures, modes of operation, etc. – can help or hinder in protecting this precious capacity.

To explore the complex interplay of person and setting, the authors employ a frame of reference pioneered at the Tavistock Clinic that strives to integrate systems thinking with psychoanalysis. Its focus renders the impact of individually experienced anxiety, guilt and doubt visible in the collective life and work of human service organizations. It also illustrates how different levels of caring systems – from the individual, to the primary work group, to the whole organization and the wider environment – interpenetrate, and how dynamics at one level can affect and be mirrored at other levels. The authors offer an enormously rich and enriching approach to understanding the powerful forces that suffuse human service organizations. I am confident that these reflections will be valuable for any managers, practitioners or students of human service organizations willing to grapple with these ideas.

I say 'grapple' for two reasons. One is that the authors have not watered down the ideas so as to make them effortlessly accessible or superficial. They are complicated ideas about complicated realities, and while the chapters are direct and understandable, they are far from simplistic. Another reason they require 'grappling' is the fullness and openness with which the authors confront the pain and despair that often permeates human service work, and that is often avoided at the cost of diminished effectiveness.

The insights in these chapters were not earned painlessly. I was impressed and, at times, humbled by the extent to which the authors made themselves vulnerable to the most primitive and disturbing aspects of the institutions they were attempting to understand and help. It is through the medium of these challenging personal experiences that the authors were often able to make sense of the dynamics that engendered symptoms of organizational dysfunction, conflict and alienation in the institutions they were attempting to assist. In reading these chapters one confronts, at arm's length, the great emotional impact of working closely with these organizations.

In this way I see the authors 'practising what they preach', so to speak. Just as the consultancies described endeavour to help caring professionals make direct, full human contact with their clients, the authors exemplify the same stance in their consultancies by openly facing and experiencing the same painful realities that were so disturbing to clients and workers in their host organizations.

From these efforts one can begin to see elements of a theory of intervention emerge, of which one component is an emphasis on devoting organizational resources to establishing forums that enable staff groups to articulate and understand their emotional experiences. The idea that through working together to understand their experiences service providers can better 'contain' and learn more deeply from them, and then in turn foster their own effectiveness, seems borne out not only by the consultancies described, but also by the evidence arising from the Consulting to Institutions Workshop in the form of this book.

A foreword would not be complete without recognizing the timeliness of this volume. Human service organizations in the United States as well as in the United Kingdom seem caught at the juncture of several powerful forces that, taken together, put them in a precarious position. Turbulent social processes are increasing the demands placed on human service professionals, often with fewer resources available to meet these needs. Accelerating rates of change in social, economic and community life are producing an ever greater sense of dislocation and anomie, and are leading to the loss of stable support systems for people and their families.

Also, among the most powerful forces affecting these institutions is the increasing reliance on markets to integrate and co-ordinate the delivery of human services. Where previously institutions assigned to meet important dependency needs of individuals were largely buffered from market forces, now they are increasingly exposed to the dictates of 'bottom line' accountability. Whether this represents progress or a widespread disavowal of social responsibility is debatable. What is unquestionable is that this shift has a profound impact on those working in these institutions, including a great increase in uncertainty and insecurity.

Combining these emerging strains with the already existing stresses of caring for people in great need amounts to a major challenge for caring professionals and their institutions, and it exacerbates their vulnerability to the powerful emotional dynamics characteristic of such work. Strategies for coping and adaptation – on both the individual and organizational levels – must be strengthened and developed. This volume is a valuable addition to our thinking about and understanding of these issues. It is a welcome contribution to much-needed efforts to help caring institutions remain vital, and to enable their staff to work with competence and compassion in the face of trying and often dispiriting conditions.

James Krantz, Ph.D.
Associate Professor of Management
Yale School of Organization and Management
and Assistant Director
Leadership Programme
Wharton School of Business
University of Pennsylvania

Preface

The global political climate as we move towards the twenty-first century is placing increasing emphasis on efficiency, cost-effectiveness, value for money – there are a whole variety of phrases in use nationally and internationally – but they all mean one thing: increased pressure on staff and on institutions. At the same time the clients, patients and other consumers of the human services, as well as the public at large, are under increasing stress themselves arising from economic, societal and geo-political pressures.

Understandably these processes have led to an increased emphasis on management and on organizational structure. Competent management carried out with clearly designated tasks and roles and backed by adequate resources can go a long way towards ensuring a creative institutional work climate. It is, however, our experience that even in the best-run institutions, not only within the human services but also in other sectors, there are pockets of irrationality and behaviour which undermine the work. There are also many institutions that are so infested with these 'anti-task' processes that the entire institution falls prey to them.

This book is an attempt to address this darker side of institutional functioning. It arises from the work of the Consulting to Institutions Workshop, started at the Tavistock Clinic in 1980, and it draws on many decades of experience in the Tavistock Clinic, the Tavistock Institute of Human Relations and in related organizations. The model used combines insights and theories arising from psychoanalysis, from systems and socio-technical approaches, from the work of Bion on groups, from Kleinian theory as applied to groups and institutions, and from group relations training originating from Kurt Lewin's work at the National Institute for Mental Health in the United States and subsequently developed by Ken Rice and others in Britain and internationally.

This book came out of the experiences of workshop members struggling not only to understand the underlying processes at work in 'people institutions', but also to make their evolving understanding available to their clients in a useful way. Although the workshop focuses on the role of the external consultant to such institutions, it seemed to us that what was needed was not a book for consultants, but an accessible book for staff and managers working in the human services, and at times caught up in institutional processes in ways which are harmful both to themselves and to the work. In this book, therefore, these processes are explored from a variety of points of view, and in a range of workplaces, so that readers have the possibility of casting beams from several perspectives to throw some light on their own experiences.

In some cases, the structures of the organizations described are changing so rapidly that the case illustrations may already seem dated. Yet, the fundamental dynamics, arising from the nature of the work itself, remain unchanged. The book is arranged to cover a mix of theories, institutions and ways of working, so that although some chapters might be more obviously relevant than others, we hope that all contributions will be of interest to readers and applicable to their work-settings.

Anton Obholzer
Vega Zagier Roberts
December 1993

Acknowledgements

The title 'The Unconscious at Work' was coined and first employed by Jon Stokes some years ago – we are indebted to him for its use. This book could not have been written without the constant and cheerful help of Sandra Masterson, who was responsible for typing and re-typing the manuscript, Vicky Davenport, who helped edit the work towards coherence, and Don Zagier and Jennifer Hicks, whose comments on successive drafts helped us keep our intended readership clearly in mind. Their support and enthusiasm for the project were invaluable. We would also like to thank Lotte Higginson, Ruth Sonntag, Lyndsay MacDonald, Margaret Walker and the staff of the Tavistock Joint Library, all members of the Consulting to Institutions Workshop both past and present, and many others.

A note on confidentiality

The examples used in this book have been selected because they are typical of situations which recur throughout the human services. Readers may therefore recognize – or think they recognize – some of the individuals and organizations described. However, all names are fictitious, and identifying details have been altered to preserve the anonymity of those involved.

Introduction
The institutional roots of consulting to institutions

James Mosse

Most people spend their working lives as part of a group which is itself part of a larger institution or organization. In this book we explore the implications of thinking about such groupings as having not only directly observable structures and functions, but also an unconscious life comparable to that described by psychoanalysis in an individual. We suggest that institutions pursue unconscious tasks alongside their conscious ones, and these affect both their efficiency and the degree of stress experienced by staff.

Since 1980, members of the Consulting to Institutions Workshop have been meeting at the Tavistock Clinic to explore the interaction between conscious and unconscious dynamics in a wide range of institutions, mainly in the statutory or voluntary care-giving sector. Their consultancy to such organizations has been directed both at fostering recognition of unconscious forces and at mitigating the negative effects of such forces. The model of consultation that is used views institutions as social systems to be studied using the established methodologies of the social sciences, but with an unconscious life to be studied psycho-analytically. The chapters that make up this book have been written by past and present members of the workshop, and describe both the theoretical roots and practical applications of their consultation work.

It is one of our themes that the social and the psychoanalytic perspectives must be deployed *together* if real change is to be effected in those aspects where structure and unconscious function overlap. Working only from the psycho-analytic perspective may heighten people's awareness of and sensitivity to unconscious processes, but will not create the conditions in which such awareness can be used, and staff will become even more depressed and frustrated. Conversely, if only the social perspective is employed, a two-dimensional blueprint for structural change may be produced, but because no account is taken of the psychic determinants of the pre-existing organization, unconscious needs are unlikely to be met by the proposed new structure, and so it will probably fail.

HISTORICAL OVERVIEW

While linking of the social and the unconscious has a history almost as long as that of the psychoanalytic conception of the unconscious itself, the thinking and manner of work described here is derived from traditions rooted in the Tavistock Institute of Human Relations. These in turn grew from the work of the Tavistock Institute of Medical Psychology (better known as the Tavistock Clinic). We start, therefore, with an historical overview of the development of these two closely linked institutions and their intellectual traditions.

The Tavistock Clinic

The Tavistock Clinic was founded in 1920 by a number of professionals, who voluntarily gave at least six hours a week to pursue psychodynamic treatments reflecting their belief that the neurotic disorders labelled as 'shell-shock', that they had been treating in combatants during the First World War, were not merely transitory phenomena related to the peculiar stresses of war, but were now endemic and pervasive in modern society. From the beginning, the aims of the Clinic were fourfold: to offer *treatment*, partly as a means of *research* into possible social means of *prevention* of such difficulties, and then to *teach* their emerging skills to other professionals. Not only did the medical staff of this founding group include general physicians, neurologists and psychiatrists, but from the first there were also psychologists and social workers. Furthermore, some of these were also trained anthropologists, so there was always a combination of a medical with a social science perspective.

In 1938, the then Medical Director of the Clinic, J. R. Rees, was appointed Consulting Psychiatrist to the Army. The army was interested not only in curing individuals, but also in developing an understanding of and treatment for the stresses of military life in time of war. During the Second World War, thirty-one ex-Tavistock staff members entered the army, many of them as staff officers in influential positions, charged with developing particular aspects of this programme. The resulting innovations can be classified as follows:

- Command Psychiatry: a reconnaissance process by senior psychiatrists who were engaged full-time in trying to identify critical problems.
- Social Psychiatry: a policy science seeking to develop preventive interventions in large-scale social problems.
- Cultural Psychiatry: a means of profiling and analysing the mentality of very large groups (initially the 'enemy nations').
- The Therapeutic Community as a means of psychiatric treatment in groups.
- The development with the military of new institutions to effect the policies derived from social psychiatry. The War Office Selection Boards were one instance of such practical developments.

The immediate postwar period

By the end of the war, a substantial number of the psychiatrists and social scientists who had been involved in the development of these innovative applications of social psychiatry to military life were determined to pursue their relevance to the postwar civilian world. A number of them returned to the Tavistock Clinic with the aim of using it as a vehicle to continue this kind of work. There followed detailed planning, intended to prepare the Tavistock to undertake its new self-appointed role. The documents produced included a discussion of 'The Place of the Tavistock Clinic in British Medicine and Social Science', the first item of which was 'The Integration of Social Science with Dynamic Psychology – a) at the level of interpersonal relations between workers, b) at the conceptual level, c) in methodology'. There were numerous references to a new discipline called 'sociatric work', and to the Tavistock's role in training people for it (Dicks 1970).

A substantial social agenda was being proposed, which was felt to require the transformation of the large prewar voluntary and part-time staff group into a smaller core of salaried, full- or almost full-time staff members. Some existing staff were asked to leave, and other new members were recruited. Criteria included commitment to the redefined social mission, but also a willingness to undertake a personal psychoanalysis, if this had not already been done; there was a perception that psychoanalytic object relations theory had proved to be relevant in the social as well as the clinical field during the war. This element in the training of the new Tavistock staff was the subject of a special arrangement with the British Psycho-Analytical Society (BPS), and reflected the BPS's part in the widespread optimism and commitment to social change – 'winning the peace' – that was already driving the formation of the reborn Tavistock, and that had swept a Labour government to power in 1945.

The Clinic explored a number of possible formal links with other institutions, both academic and clinical, which they hoped might further their agenda. For a variety of reasons, however, not much came of these. Meanwhile, the emergence of plans for the new National Health Service brought a clear problem. The therapeutic work could logically become a part of the NHS, but the research, development and training activities envisaged as part of the heightened social engagement stood outside the scope of the NHS. Indeed, once within it, the Clinic would be precluded from accepting many of the projects and sources of funding it was so keen to pursue. Preparations were therefore made for the incorporation as a separate non-profit organization of the Tavistock Institute of Human Relations. This took place in September 1947, leaving the Clinic free to enter the NHS in July 1948. As a part of the same strategic manoeuvring to ensure the long-term security of the social agenda, a new international learned journal, *Human Relations*, was launched jointly with Kurt Lewin's Research Center for Group Dynamics in the United States, and a new publishing imprint, Tavistock Publications, was established.

The Tavistock Institute of Human Relations (TIHR)

By 1948, the British government was sufficiently worried about the low levels of productivity in the postwar economy to establish an Industrial Productivity Committee. This included a Human Factors Panel, which was empowered to grant-aid research intended to improve productivity through better use of human resources. The fledgeling TIHR applied for and secured funding for three projects. One was a training project for a group of six 'industrial fellows', to be seconded from industry for two years to learn about unconscious group processes in the workplace from their own experience. Each of them would take part in other TIHR research projects, and also participate in a therapy group to study their own unconscious group life. They would then return to their seconding organizations, and, by applying their newly gained experience, would help to change work practices in beneficial ways. This can be seen as one of the forerunners of the Group Relations Training Programme (see Chapter 4).

Another early project to receive a grant was a study by Eric Trist of work organization in the newly nationalized coal-mining industry. It was discovered that groups of workers supposedly doing similar jobs in separate coal mines in fact organized themselves very differently, and that this had significant effects on levels of productivity. This led to the concept of the self-regulating work group, and to the idea that differences in group organization reflect unconscious motives, which also affect the subjective experience of the work. It was through this project that the 'socio-technical system' came to be defined as an appropriate field of study (Trist *et al.*, 1963). Organizations as socio-technical systems can be understood as the product of the interaction between a work task, its appropriate techniques and technology, and the social organization of the workers pursuing it. While originating from research in industry, this approach has subsequently been applied to the study of a wide range of organizations. In particular, Isabel Menzies' (1960) study to identify the causes of the high drop-out rate from nurse training was an early example of bringing the TIHR socio-technical model to bear on an institution where the technical system is largely human.

The last of these three original projects was a detailed study of the internal relations of a single manufacturing company. The aim was to identify ways of improving co-operation between all levels of staff, and then help the company to move towards implementing these. The project was undertaken by Elliott Jaques at the Glacier Metal Company, and led to extensive change in the culture and organization of the firm. However, though it was widely reported and studied, it was not much copied. One of Jaques' most significant contributions resulting from this project was the recognition that social systems in the workplace function to defend workers against unconscious anxieties inherent in the work. To the extent that such defences are unconscious, the social systems are likely to be rigid and therefore uncomfortable; but because of their role in keeping anxiety at bay, they may also be very resistant to change (Jaques 1951, 1953).

Most of the core theoretical influences on the work discussed in this book derive from these first three projects. The one major addition still to be introduced is that of open systems. From the first, systems theory was one of the imports from the social sciences that underpinned socio-psychological thinking. The particular application of open systems theory to the work of TIHR was substantially the contribution of A. K. Rice, later working together with Eric Miller. In essence, the open systems view sees an institution as having boundaries across which inputs are drawn in, processed in accordance with a primary task, and then passed out as outputs. While this may sound like a model best suited to understanding manufacturing processes, Miller and Rice (1967) applied it far more widely. They traced many of the difficulties faced by work groups to their problems in defining their primary task and in managing their boundaries (see Chapter 3).

TIHR personnel did not go in as experts who already knew what their clients must do to improve things: they went to study whatever they would find. The study was undertaken jointly *with* the clients, and, to a large extent, *by* them. TIHR staff then sought to contribute a way of construing their observations and experiences, which they believed would point to potentially helpful changes. Once introduced, the effects of the changes would themselves become the subject of further study, leading to further change. The role of the TIHR staff member was designated as 'participant observer', and the whole style of working was known as 'action research'. This was a continuation of the Clinic's longstanding doctrine, 'No research without therapy, and no therapy without research' – though now applied to institutions rather than individuals (Dicks 1970).

THE TOOLS OF CONSULTANCY

As government and charitable funding became more difficult to secure through the 1950s and 1960s, the TIHR shifted its focus from grant-aided research towards consultancy work directly commissioned by client organizations. Since then, TIHR staff have undertaken a large number of both research and consultancy projects combining perspectives both from the social sciences and from psychoanalysis, a body of theory and practice which has shaped many of the interventions described in this book. It has also influenced the training and consultancy activities of many organizations worldwide (see Appendix, p. 211). Together, these two perspectives provide us with tools not only for taking up a role as consultant, but also for thinking about our experiences as members of institutions.

From the social sciences

A consultant engaging with a client organization is engaging with a social system. This system exists in the real world, and has a structure intended to relate to the effective discharge of its primary task. Both this structure and the technology of task performance must therefore be understood, as must the interface between them. This means understanding the organization's description of itself and its intended structure. The consultant, however, also must be able to observe for him- or herself what actually goes on, regardless of what is claimed, and then be able to reflect upon the significance of what has been discerned.

Exploring social structures has long been the province of the social sciences. Some of the contributors to this book have studied anthropology or sociology; others have gained their experience in thinking formally about social systems through membership of a Tavistock group relations training conference (see Chapter 4). In addition, we all have wide and varied experience of institutional membership derived from both working and private life. Families, schools, clubs, professional bodies and employing organizations are all institutions in this sense. A capacity to keep all these experiences alive and available for recall has many benefits. This includes helping one to stay on the margin of a group, rather than being sucked into shared assumptions about how things 'have' to be organized.

For example, I have worked both in industry and in a multidisciplinary mental health team, and thus have experienced two very different sets of beliefs about delegation and decision-making. In industry, I heard suggestions for widespread consultation before reaching a decision described as 'analysis paralysis'. In the health service, on the other hand, I have seen a capacity for effective action regarded as 'manic', and detailed planning ridiculed as 'obsessional'. My own efforts to reconcile these differing views of how decisions 'must' be taken have been of considerable value in a wide range of situations, not least my consultancy work. Being able to call on a diverse range of experience can also reduce the likelihood of being swept away by the institutional defences of a group of which one is a part by ensuring that no single role is too easily mobilized in isolation. For the consultant, it also reduces the risk of being seduced into the role of expert, manager or supervisor.

Social science aims to relate observable social structures to their functions in the external world. These are held to be, at least in principle, directly accessible to consciousness. Any institution having sufficient coherence and money to hire a consultant will support such an analysis. On its own, however, this can produce a two-dimensional blueprint rather than a three-dimensional working model. The so-described inanimate institution is also made up of living people who have unconscious and non-rational aims and needs, which they must serve simultaneously with the rational aims of the institution. The pure social scientist has neither the tools nor the theoretical framework for observing this third dimension.

From psychoanalysis

These tools and theories can be drawn from psychoanalysis. Freud writes that the analyst 'must turn his own unconscious like a receptive organ towards the transmitting unconscious of the patient' (1924: 115). It is axiomatic, and stands at the very heart of applied psychoanalytic work, that the instrument with which one explores unconscious processes is oneself – one's own experience of and feelings about the shared situation. If the self is to be the scientific instrument on which 'readings' are taken, then how is this instrument to be calibrated? The answer from psychoanalysis is unequivocal: through personal analysis. Anyone wishing to work as an analyst must first undertake an analysis of their own, through which they should be able to distinguish what comes from themselves – their own unresolved conflicts – and what belongs to the patient. They should also gain experientially based understanding of theoretical concepts described in the literature. In order to undertake the kind of institutional consultation described in this book, *some* personal therapy is probably necessary, sufficient to help one to 'catch' and reorient oneself within the powerful unconscious psychic currents that run through groups, particularly when their unconscious defences are under scrutiny.

There are two main dangers in seeking to apply a purely psychoanalytic perspective to institutions. The first is that it may lead to attempts to develop members' 'sensitivity' and insight into their own and the institution's psychological processes, while ignoring the systemic elements that affect the work. In this case, instead of bringing about useful and needed change, their heightened sensitivity may add to members' frustration and have a negative effect on the institution (Menzies Lyth 1990). The second is the risk of what has been called 'character assassination', in which psychoanalytic theory is misused to disparage character and impugn motives. This can lead to attributing institutional problems to the individual pathology of one or more of its members (see Chapter 14). It can also lead to consultants' undertaking or presenting their work in a way that pathologizes the behaviour and functioning of the institution and its individual members without giving due regard to the effectiveness with which the conscious real-world tasks of the organization are being pursued. For example, a staff group may be efficiently producing an award-winning newspaper as well as experiencing sufficient difficulties in the work to seek help from an institutional consultant. The real-world success of the institution, however, may be undervalued or ignored by a consultant whose focus is almost entirely on the behaviour and experiences of individuals. This difficulty may arise innocently, as an empathic response to the observable suffering of the staff; it can also result from a tendency to disparage and discount the expertise of others. This is another area in which continuing access to a wide range of membership of different institutions may be an invaluable bulwark against being swept away by the shared assumptions of any one group, either those of the client institution, or of one's home-group whose assumptions the client is most powerfully challenging.

Maintaining an outside perspective

We all participate as members in a large number of different institutions, of which our workplace is usually one of the most structured and demanding. Often we are acutely aware that our involvement is causing us distress, and yet for all our cries of 'If only *that* was different!' we cannot put our finger on what is wrong. For some reason, it seems terribly hard to use our powers of observation, or our memories of how things were elsewhere, to bring about change. What prevents us from doing this is the main theme of this book. Our central hypothesis is that membership of an institution makes it harder to observe or understand that institution: we become caught up in the anxieties inherent in the work and the characteristic institutional defences against those anxieties. This soon leads to shared, habitual ways of seeing, and a common failure to question 'holy writ'. Newcomers may be able to see more clearly, but have no licence to comment. By the time they do, they have either forgotten how to see, or have learned not to. They, too, require defending against anxieties, not least the anxiety of upsetting their colleagues.

By definition, consultants have a stance outside the daily life of the institution. This makes it easier for them to observe, and to think about what they observe, without getting caught up in institutional defences. It may also make it easier for the institution to license them to see and be heard. Consultants also have at their disposal particular tools to make sense of what they see and feel. It is our tool-kit that we offer in this book. The 'tools' are explained in Part I; the other parts illustrate how they can be used. The tool-kit does not purport to be complete; other tools from other kits will be more appropriate in certain situations. Our hope is to offer readers some ways of thinking and managing themselves that may enable them to function more effectively and with less distress. We are not experts on how to do their work. Rather, we hope *our* skills lie in helping to liberate *their* expertise.

Part I

Conceptual framework

INTRODUCTION

When people approach the Tavistock for help with work-related difficulties, or for training in understanding organizational processes, they often have some idea that there is a 'Tavistock model'. To a large extent, this is a myth. Indeed, the Tavistock itself consists of several parts which are more or less independent of each other. Of these, the Tavistock Clinic (itself comprising several departments), the Tavistock Institute of Human Relations and the Tavistock Marital Studies Institute all offer different kinds of organizational consultancy and training, based on different, though overlapping, theories. Even within a single department, the theory informing the staff's thinking and approach will vary. And yet there are ways of trying to understand what goes on in organizations and in the individuals who work in them which could be regarded as 'more Tavistock than otherwise'.

The contributors to this book draw on a number of interwoven strands of theory which together provide the conceptual framework on which their work is based. Historically, this can be traced back to the convergence of two previously distinct intellectual traditions – psychoanalysis and social science – which took place at the Tavistock Institute of Human Relations after World War II. The development of this framework is discussed in the introduction to the book, which also presents the argument for drawing on both these traditions, and indicates some of the risks in neglecting one or the other. However, in this part the main strands are 'unpicked' and discussed separately so that essential ideas can be presented as clearly as possible.

The first chapter introduces psychoanalytic concepts which can be of help in making sense of seemingly irrational processes in individuals, groups and organizations. The second chapter discusses the work of the psychoanalyst Wilfred Bion on groups. Bion's theory of group behaviour is widely known and often quoted, though sometimes in ways that do little to add to real understanding of unconscious group processes. This chapter, in addition to summarizing the theory, also illustrates how it can be used to shed light on some of the frustrating and puzzling experiences we all have in committees, teams and other work groups. The third chapter introduces concepts from open systems theory. These provide a model for examining how work is organized and how the resulting structures are managed.

In addition to the theories outlined in the first three chapters, most of the authors of this book have been greatly influenced by their experiences both as members and as staff of group relations conferences. These conferences, first run by the Tavistock Institute of Human Relations in 1957, have regularly focused on the theme of authority and leadership, from which both the title of many of the conferences and the title of the last chapter in this section have been drawn. An understanding of authority, power and leadership is essential to understanding organizational functioning; it also illuminates some of the difficulties we have in managing ourselves at work, as well as in being managed and managing others.

Together, these four chapters present the theoretical underpinnings of the chapters which follow. Some readers may prefer to skip them initially, returning to them for reference when they encounter the concepts later in the book and become curious about them.

Chapter 1

Some unconscious aspects of organizational life

Contributions from psychoanalysis

William Halton

'None of your jokes today,' said an eight-year-old coming into the consulting room. Interpretations about the process of the unconscious may often seem like bad jokes to the recipient, if not frankly offensive. Although some elements of psychoanalysis have become part of everyday life, psychoanalysis as a treatment for individual emotional problems remains a minority experience. As a system of ideas it has adherents, sceptics and a multitude of indifferent passers-by.

Despite the fact that there is no exact parallel between individuals and institutions, psychoanalysis has contributed one way to approach thinking about what goes on in institutions. This approach does not claim to provide a comprehensive explanation or even a complete description. But looking at an institution through the spectrum of psychoanalytical concepts is a potentially creative activity which may help in understanding and dealing with certain issues. The psychoanalytical approach to consultation is not easy to describe. It involves understanding ideas developed in the context of individual therapy, as well as looking at institutions in terms of unconscious emotional processes. This may seem like a combination of the implausible with the even more implausible or it may become an illuminating juxtaposition.

THE UNCONSCIOUS

As Freud and others discovered, there are hidden aspects of human mental life which, while remaining hidden, nevertheless influence conscious processes. In treating individuals, Freud found that there was often resistance to accepting the existence of the unconscious. However, he believed he could demonstrate its existence by drawing attention to dreams, slips of tongue, mistakes and so forth as evidence of meaningful mental life of which we are not aware. What was then required was interpretation of these symbolic expressions from the unconscious.

Ideas which have a valid meaning at the conscious level may at the same time carry an unconscious hidden meaning. For example, a staff group talking about their problems with the breakdown of the switchboard may at the same time be making an unconscious reference to a breakdown in interdepartmental communication. Or complaints about the distribution of car-park spaces may also be a symbolic communication about managers who have no room for staff concerns. The psychoanalytically oriented consultant takes up a listening position on the boundary between conscious and unconscious meanings, and works simultaneously with problems at both levels. It may be some time before the consultant can pick up and make sense of these hidden references to issues of which the group itself is not aware.

THE AVOIDANCE OF PAIN

Like individuals, institutions develop defences against difficult emotions which are too threatening or too painful to acknowledge. These emotions may be a response to external threats such as government policy or social change. They may arise from internal conflicts between management and employees or between groups and departments in competition for resources. They may also arise from the nature of the work and the particular client group, as described in detail in Part II of this book. Some institutional defences are healthy, in the sense that they enable the staff to cope with stress and develop through their work in the organization. But some institutional defences, like some individual defences, can obstruct contact with reality and in this way damage the staff and hinder the organization in fulfilling its task and in adapting to changing circumstances. Central among these defences is *denial*, which involves pushing certain thoughts, feelings and experiences out of conscious awareness because they have become too anxiety-provoking.

Institutions call in consultants when they can no longer solve problems. The consultant who undertakes to explore the nature of the underlying difficulty is likely to be seen as an object of both hope and fear. The conscious hope is that the problem will be brought to the surface, but at the same time, unconsciously, this is the very thing which institutions fear. As a result, the consultant's interpretations of the underlying unconscious processes may well meet with *resistance*, that is, an emotionally charged refusal to accept or even to hear what he or she says. The consultant may only gradually be able to evaluate the nature of the defences, reserving interpretation until the group is ready to face what it has been avoiding and to make use of the interpretation.

A symbolic communication may occur just at a point where the consultant's understanding of the hidden meaning coincides with the group's readiness to receive it. The following example illustrates a symbolic communication occurring in a group unable to accept the reality of a threat from an external source:

The Manfred Eating Disorders Unit at Storsey Hospital was conducting a last-ditch campaign against closure, and had engaged the help of a consultant. The campaign had been running for several months, with the staff working late and over weekends organizing a local petition and lobbying professional colleagues, local councillors and MPs. Public opinion was on the side of keeping the unit open; closure seemed unthinkable. New patients were still being admitted and no preparation had been made for transferring existing patients. Only the consultant did not share the excited mood and felt depressed after each meeting.

At one meeting the discussion strayed to the apparently irrelevant topic of the merits of euthanasia. The consultant heard this as an indirect expression of the anticipated relief which the completion of closure would bring, and said this to the group. Their first reaction was one of shock. However, over time the interpretation led to a more realistic attitude to the possibility of closure, which made it possible for the staff to begin to think and plan for the patients.

In this example, the staff had been responding to the threat of closure with angry and excited activity aimed at saving the unit. By relating only to the possibility of winning their fight for survival, they resisted and denied the possibility of closure, ignoring the reality of financial cuts which had already closed other local units. The motivation behind this resistance was not only the wish to save the unit for the benefit of patients; closure would also hurt their pride, cast doubts on the value of their work, and cause other emotional pain. But insofar as they had lost touch with reality, they had also failed in their responsibility to prepare patients for the possibility of closure. As an outsider, the consultant felt the depression which was so strikingly absent in the group; this was an important clue to what was being avoided. The euthanasia discussion indicated that previously denied feelings were moving towards the surface as a symbolic communication, and that the group was ready to acknowledge them.

THE CONTRIBUTION OF MELANIE KLEIN

In play, children represent their different feelings through characters and animals either invented or derived from children's stories: the good fairy, the wicked witch, the jealous sister, the sly fox and so on. This process of dividing feelings into differentiated elements is called *splitting*. By splitting emotions, children gain relief from internal conflicts. The painful conflict between love and hate for the mother, for instance, can be relieved by splitting the mother-image into a good fairy and a bad witch. *Projection* often accompanies splitting, and involves locating feelings in others rather than in oneself. Thus the child attributes slyness to the fox or jealousy to the bad sister. Through play, these contradictory feelings and figures can be explored and resolved.

Through her psychoanalytic work with children in the early 1920s, Melanie Klein developed a conceptualization of an unconscious inner world, present in everyone, peopled by different characters personifying differentiated parts of self or aspects of the external world. Early in childhood, splitting and projection are the predominant defences for avoiding pain; Klein referred to this as the *paranoid-schizoid position* ('paranoid' referring to badness being experienced as coming from outside oneself, and 'schizoid' referring to splitting). This is a normal stage of development; it occurs in early childhood and as a state of mind it can recur throughout life. Through play, normal maturation or psychoanalytic treatment, previously separated feelings such as love and hate, hope and despair, sadness and joy, acceptance and rejection can eventually be brought together into a more integrated whole. This stage of integration Klein called the *depressive position*, because giving up the comforting simplicity of self-idealization, and facing the complexity of internal and external reality, inevitably stirs up painful feelings of guilt, concern and sadness. These feelings give rise to a desire to make reparation for injuries caused through previous hatred and aggression. This desire stimulates work and creativity, and is often one of the factors which leads to becoming a 'helping' professional (see also Chapter 12).

THE PARANOID-SCHIZOID POSITION

The discovery from child analysis that the different and possibly conflicting emotional aspects of an experience may be represented by different people or different 'characters' is used in institutional consultancy as a guide for understanding group processes. In play, the child is the originator of the projections and the play-figures are the recipients. In an institution, the client group can be regarded as the originator of projections with the staff group as the recipients. The staff members may come to represent different, and possibly conflicting, emotional aspects of the psychological state of the client group. For example, in an adolescent unit the different and possibly conflicting needs of the adolescent may be projected into different staff members. One member may come to represent the adolescent's need for independence while another may represent the need for limits. In an abortion clinic, one nurse may be in touch only with a mother's mourning for her lost baby, while another may be in touch only with the mother's relief. These projective processes serve the same purpose for the client as play does for the child: relief from the anxieties which can arise from trying to contain conflicting needs and conflicting emotions. It is hard to contain mourning and relief simultaneously, or to experience the wish for independence and the need for limits at the same time. The splitting and projection of these conflicting emotions into different members of the staff group is an inevitable part of institutional process.

Schizoid splitting is normally associated with the splitting off and projecting outwards of parts of the self perceived as bad, thereby creating external figures who are both hated and feared. In the helping professions, there is a tendency to deny feelings of hatred or rejection towards clients. These feelings may be more easily dealt with by projecting them onto other groups or outside agencies, who can then be criticized. The projection of feelings of badness outside the self helps to produce a state of illusory goodness and self-idealization. This black-and-white mentality simplifies complex issues and may produce a rigid culture in which growth is inhibited.

A student occupation protesting against the effects of government cuts was condemned by the college authorities as a ludicrous and irrational waste of resources. They were quoted in the press as saying: 'The question remains as to who was leading this disruption. We have had complaints from students suggesting that outsiders and non-students took active parts in proceedings in meetings and occupation action . . . there was no apparent agenda other than disruption.'

The implication was that inside the college there were good students and good managers. Outside there were bad people who contaminated and disrupted the institution, manipulating its members for destructive purposes. Splitting and projection exploits the natural boundary between insiders and outsiders which every institution has. In this example it led to a state of fragmentation because contact was lost between parts of the institution which belonged together inside its boundary. There was no dialogue possible between the conflicting points of view within the college, and so change and development were frustrated.

Sometimes the splitting process occurs between groups within the institution. Structural divisions into sections, departments, professions, disciplines and so forth are necessary for organizations to function effectively. However, these divisions become fertile ground for the splitting and projection of negative images. The gaps between departments or professions are available to be filled with many different emotions – denigration, competition, hatred, prejudice, para- noia. Each group feels that it represents something good and that other groups represent something inferior. Doctors are authoritarian, social workers talk too much, psychotherapists are precious, managers only think about money. Indivi- dual members of these groups are stereotyped like the characters playing these roles in children's games and stories. The less contact there is with other sections, the greater the scope for projection of this kind. Contact and meetings may be avoided in order unconsciously to preserve self-idealization based on these projections. This results in the institution becoming stuck in a paranoid-schizoid projective system. Emotional disorder interferes with the functioning of an organization, particularly in relation to tasks which require co-operation or collective change. (For further examples of these processes, see Chapters 8, 9, 18 and 20.)

ENVY

On occasion, difficulties in collaboration arise not so much from the desire to be an ideal carer or a more potent worker, but from a sense of being an inevitable loser in a competitive struggle. In the current climate of market values and shrinking budgets, the success of one part of the organization can be felt to be at the expense of another. The survival-anxiety of the less successful section stimulates an envious desire to spoil the other's success. This spoiling envy operates like a hidden spanner-in-the-works, either by withholding necessary co-operation or by active sabotage.

A group of lecturers at Branston Polytechnic were organizing a money-raising short course out of term-time. They found it impossible to gain the co-operation of the catering department, who refused to provide tea and coffee for course participants. Then, on the first day of the course, despite the lecturers' having requested a delay, maintenance work was started on the toilet facilities, putting them out of action. Both the catering and maintenance departments were about to be closed down, their services to be taken over by private organizations following tendering. Their unco-operativeness led to considerable inconvenience for the participants, who blamed the course organizers and left the course very critical of the polytechnic as a whole.

Catering and maintenance staff felt devalued by the decision to 'sell off' their services, and envious of the academic staff's protected status. This kind of spoiling envy often gives rise to hostile splits between parts of an organization such that the enterprise as a whole is damaged.

PROJECTIVE IDENTIFICATION AND COUNTERTRANSFERENCE

Although psychoanalysis is based on the idea that the behaviour of an individual is influenced by unconscious factors, the psychoanalytic view of institutional functioning regards an individual's personal unconscious as playing only a subsidiary role. Within organizations, it is often easier to ascribe a staff member's behaviour to personal problems than it is to discover the link with institutional dynamics. This link can be made using the psychoanalytic concept of *projective identification*. This term refers to an unconscious inter-personal interaction in which the recipients of a projection react to it in such a way that their own feelings are affected: they unconsciously identify with the projected feelings. For example, when the staff of the Manfred Eating Disorders Unit projected their depression about closure into the consultant, he felt this depression as if it were his own. The state of mind in which other people's feelings are experienced as one's own is called the *countertransference*.

Projective identification frequently leads to the recipient's acting out the countertransference deriving from the projected feelings. For example, the staff of an adolescent unit may begin to relate to each other as if they were adolescents themselves, or may act in adolescent ways such as breaking the rules and otherwise challenging authority figures. Such behaviour indicates that projective identification is at work, but the true source of their feelings and behaviour is likely to remain obscure until staff achieve a conscious realization that they have become trapped in a countertransference response to a projective process. (This is discussed in more detail in Chapter 5, and further illustrated in Chapter 7.)

It is also through the mechanism of projective identification that one group on behalf of another group, or one member of a group on behalf of the other members, can come to serve as a kind of 'sponge' for all the anger or all the depression or all the guilt in the staff group. The angry member may then be launched at management by the group, or a depressed member may be unconsciously manoeuvred into breaking down and leaving. This individual not only expresses or carries something for the group, but may be used to export something which the rest of the group then need not feel in themselves (see Chapter 14). Similarly, a group may carry something for another group or for the institution as a whole (see Chapter 8). If there is something which a group cannot bear at all, like the depression about the closure of the Manfred Unit, it may call in a consultant to carry that feeling on its behalf.

THE DEPRESSIVE POSITION

When we recognize that our painful feelings come from projections, it is a natural response to 'return' these feelings to their source: 'These are *your* feelings, not mine.' This readily gives rise to blaming, and contributes to the ricocheting of projections back and forth across groups and organizations. However, if we can tolerate the feelings long enough to reflect on them, and *contain* the anxieties they stir up, it may be possible to bring about change. At times when we cannot do this, another person may temporarily contain our feelings for us. This concept of a person as a 'container' comes from the psychoanalytic work of Bion (1967). He likened it to the function of the mother whose ability to receive and understand the emotional states of her baby makes them more bearable. (This is discussed further in Chapters 5 and 7.)

Certainly, both psychoanalysts and psychoanalytical consultants aim to identify the projective processes at work and trace the projections to their source, but this in itself is not enough. What was previously unbearable – and therefore projected – needs to be made bearable. It is painful for the individual or group or institution to have to take back less acceptable aspects of the self which had previously been experienced as belonging to others: for example, that legitimate criticisms may arise from within a college, and not simply as intrusions by malicious outsiders; or that no psychiatric unit is so ideal that it cannot be closed; or that in adolescents the need for independence and the need for limits are equally valid concerns and should be held in a complementary tension; or that abortion gives rise to both mourning and relief; or that good managers and good caretakers are both necessary to make a healthy organization; or that authorities who implement cuts and students who protest may both care about the education system.

The consultant's willingness and ability to contain or hold on to the projected feelings stirred up by these ambiguities until the group is ready to use an interpretation are crucial. Otherwise the interpretation will be experienced as yet another attack. However, when the timing is right, some of the projections can be 're-owned', splitting decreases, and there is a reduction in the polarization and antagonism among staff members themselves. This promotes integration and co-operation within and between groups or, in psychoanalytic terms, a shift from the paranoid-schizoid to the depressive position.

In a group functioning in the depressive position, every point of view will be valued and a full range of emotional responses will be available to it through its members. The group will be more able to encompass the emotional complexity of the work in which they all share, and no one member will be left to carry his or her fragment in isolation. Furthermore, in order to contain the tendency towards splitting in the client group, the staff group must be able to hold together the conflicting elements projected into them, discussing and thinking them through instead of being drawn into acting them out. This requires being aware of the particular stresses involved in their work, as well as recognizing its limitations. The lessening of conflict may then open the way to better working practices and greater job satisfaction, as staff process and integrate their collective work experience. However, the depressive position is never attained once and for all. Whenever survival or self-esteem are threatened, there is a tendency to return to a more paranoid-schizoid way of functioning.

CONCLUSION

Psychoanalytical concepts make a particular contribution to thinking about institutional processes, though contributions from other conceptual frameworks are also necessary to understand institutional functioning. Psychoanalysis is concerned with understanding the inner world with its dynamic processes of fragmentation and integration; key concepts include denial of internal and external reality, splitting, projection and idealization.

Psychoanalytically oriented consultants extend these concepts to understanding unconscious institutional anxieties and the defences against them. Besides concepts, they bring from psychoanalysis a certain stance or frame of mind: to search for understanding without being judgemental either of their clients or of themselves. This enables them to make themselves available to receive and process projections from the institution. The feelings experienced by the consultant or, indeed, by any member of an institution, while interacting with it, constitute the basic countertransference response on which the understanding of unconscious institutional processes is based.

At its best, such understanding can create a space in the organization in which staff members can stand back and think about the emotional processes in which they are involved in ways that reduce stress and conflict, and can inform change and development. The ideas discussed in this and subsequent chapters can be used to develop a capacity for self-consultation: for observing and reflecting on the impact unconscious group and organizational processes have on us all, and our own contribution to these processes as we take up our various roles.

Chapter 2

The unconscious at work in groups and teams

Contributions from the work of Wilfred Bion

Jon Stokes

Our experiences of being and working in groups are often powerful and overwhelming. We experience the tension between the wish to join together and the wish to be separate; between the need for togetherness and belonging and the need for an independent identity. Many of the puzzling phenomena of group life stem from this, and it is often difficult to recognize the more frequent reality of mutual interdependence. No man is an island, and yet we wish to believe we are independent of forces of which we may not be conscious, either from outside ourselves or from within. At times we are aware of these pulls within ourselves; at other times they overwhelm us and become the source of irrational group behaviour. While most obvious in crowds and large meetings, these same forces also influence smaller groups, such as teams and committees. This chapter will focus on groups at work, and how they are affected by unconscious processes.

The psychoanalytic study of unconscious processes in groups begins with Freud's *Group Psychology and the Analysis of the Ego* (1921). Essentially, Freud argued that the members of a group, particularly large groups such as crowds at political rallies, follow their leader because he or she personifies certain ideals of their own. The leader shows the group how to clarify and act on its goals. At the same time, the group members may project their own capacities for thinking, decision-making and taking authority on to the person of the leader and thereby become disabled. Rather than using their personal authority in the role of follower, the members of a group can become pathologically dependent, easily swayed one way or another by their idealization of the leader. Criticism and challenge of the leader, which are an essential part of healthy group life, become impossible. (This is further discussed in Chapter 4.)

WILFRED BION AND BASIC ASSUMPTIONS IN GROUPS

A major contributor to our understanding of unconscious processes in groups was the psychoanalyst Wilfred Bion, who made a detailed study of the processes in small groups in the army during World War II, and later at the Tavistock Clinic. On the basis of these, he developed a framework for analysing some of the more irrational features of unconscious group life. His later work on psychosis, thinking and mental development (Bion 1967, 1977) has also contributed much to our understanding of groups and organizational processes and is referred to elsewhere in this book (see Chapters 1, 7, 12 and 18). Bion himself wrote little further on groups as such, preferring to concentrate on the internal world of the individual. In fact, as he himself argues, the group and the psychoanalytic pair of psychoanalyst and analysand actually provide two different 'vertices' on human mental life and behaviour. Each is distinct but not mutually incompatible, just as, for example, physics and chemistry provide distinct levels of understanding of the material world. Indeed, the whole matter of the relationship between the individual and the group is a central theme throughout both Bion's work and his life (Armstrong 1992 and Menzies Lyth 1983 – for a further understanding of Bion's work see Anderson 1992; Meltzer 1978; and Symington 1986: Chapters 26 and 27). This chapter concentrates solely on some of the implications for understanding groups and teams based on the ideas contained in Bion's *Experiences in Groups* (1961).

Bion distinguished two main tendencies in the life of a group: the tendency towards work on the primary task (see Chapter 3) or *work-group mentality*, and a second, often unconscious, tendency to avoid work on the primary task, which he termed *basic assumption mentality*. These opposing tendencies can be thought of as the wish to face and work *with* reality, and the wish to evade it when it is painful or causes psychological conflict within or between group members.

> The staff of a day centre spent a great deal of time arguing about whether or not the clients should have access to an electric kettle to make drinks with. Some were strongly of the opinion that this was too dangerous, while others were equally adamant that the centre should provide as normal an environment as possible. While there was a real policy issue, the argument was also an expression of the difficulty the staff were having with their angry and violent feelings towards their clients, who were behaving in ways that frustrated the staff's wish that they 'get better'. The fear of the clients' scalding themselves also contained a less conscious and unspoken wish to punish them. However, it was too painful for the staff to face these feelings. Instead, each time the ostensible problem was near to solution, some new objection would be raised, with the result that the group was in danger of spending the whole of its weekly team meeting on the matter of the kettle. An interpretation of this problem by the consultant enabled a deeper discussion of the ambivalent feelings, and a return to the group's work task: the exploration of working relations and practices in the centre.

In work-group mentality, members are intent on carrying out a specifiable task and want to assess their effectiveness in doing it. By contrast, in basic assumption mentality, the group's behaviour is directed at attempting to meet the unconscious needs of its members by reducing anxiety and internal conflicts.

THE THREE BASIC ASSUMPTIONS

How groups do this varies. According to Bion, much of the irrational and apparently chaotic behaviour we see in groups can be viewed as springing from *basic assumptions* common to all their members. He distinguished three basic assumptions, each giving rise to a particular complex of feelings, thoughts and behaviour: *basic assumption dependency*, *basic assumption fight-flight* and *basic assumption pairing*.

Basic assumption dependency (baD)

A group dominated by baD behaves as if its primary task is solely to provide for the satisfaction of the needs and wishes of its members. The leader is expected to look after, protect and sustain the members of the group, to make them feel good, and not to face them with the demands of the group's real purpose. The leader serves as a focus for a pathological form of dependency which inhibits growth and development. For example, instead of addressing the difficult items on the agenda, a committee may endlessly postpone them to the next meeting. Any attempts to change the organization are resisted, since this induces a fear of being uncared for. The leader may be absent or even dead, provided the illusion that he or she contains the solution can be sustained. Debates within the organization may then be not so much about how to tackle present difficulties as about what the absent leader would have said or thought.

Basic assumption fight-flight (baF)

The assumption here is that there is a danger or 'enemy', which should either be attacked or fled from. However, as Bion puts it, the group is prepared to do either indifferently. Members look to the leader to devise some appropriate action; their task is merely to follow. For instance, instead of considering how best to organize its work, a team may spend most of the time in meetings worrying about rumours of organizational change. This provides a spurious sense of togetherness, while also serving to avoid facing the difficulties of the work itself. Alternatively, such a group may spend its time protesting angrily, without actually planning any specific action to deal with the perceived threat to its service.

Basic assumption pairing (baP)

BaP is based on the collective and unconscious belief that, whatever the actual problems and needs of the group, a future event will solve them. The group behaves as if pairing or coupling between two members within the group, or perhaps between the leader of the group and some external person, will bring about salvation. The group is focused entirely on the future, but as a defence against the difficulties of the present. As Bion puts it, there is a conviction that the coming season will be more agreeable. In the case of a work team, this may take the form of an idea that improved premises would provide an answer to the group's problems, or that all will be well after the next annual study day. The group is in fact not interested in working practically towards this future, but only in sustaining a vague sense of hope as a way out of its current difficulties. Typically, decisions are either not taken or left extremely vague. After the meeting, members are inevitably left with a sense of disappointment and failure, which is quickly superseded by a hope that the next meeting will be better.

RECOGNIZING BASIC ASSUMPTION ACTIVITY

The meetings of a group of psychologists to which I consulted would often start with a discussion of their frustration at decisions not having been implemented. At one meeting, the main topic for a considerable time was the previous meeting, whether it had been a good meeting or a bad meeting – it being entirely unclear what this meant. When I pointed this out, there followed a lengthy debate about the relative merits of various chairs, seating arrangements, and, finally, rooms in which to hold the meeting. Various improved ways of organizing the meeting were proposed, but no decision was reached. I suggested there was a fear of discussing matters of real concern to the members present, perhaps a fear of conflict. At this point it emerged that there was indeed considerable controversy about a proposed appointment, some favouring one method, others another. Eventually a decision was almost reached, only to be resisted on the grounds that one significant member of the team was absent.

When under the sway of a basic assumption, a group appears to be meeting as if for some hard-to-specify purpose upon which the members seem intently set. Group members lose their critical faculties and individual abilities, and the group as a whole has the appearance of having some ill-defined but passionately involving mission. Apparently trivial matters are discussed as if they were matters of life or death, which is how they may well feel to the members of the group, since the underlying anxieties are about psychological survival.

In this state of mind, the group seems to lose awareness of the passing of time, and is apparently willing to continue endlessly with trivial matters. On the other hand, there is little capacity to bear frustration, and quick solutions are favoured. In both cases, members have lost their capacity to stay in touch with reality and its demands. Other external realities are also ignored or denied. For example, instead of seeking information, the group closes itself off from the outside world and retreats into paranoia. A questioning attitude is impossible; any who dare to do so are regarded as either foolish, mad or heretical. A new idea or formulation which might offer a way forward is likely to be too terrifying to consider because it involves questioning cherished assumptions, and loss of the familiar and predictable which is felt to be potentially catastrophic. At the prospect of any change, the group is gripped anew by panic, and the struggle for understanding is avoided. All this prevents both adaptive processes and development (Turquet 1974). Effective work, which involves tolerating frustration, facing reality, recognizing differences among group members and learning from experience, will be seriously impeded.

LEADERSHIP AND FOLLOWERSHIP IN BASIC ASSUMPTION GROUPS

True leadership requires the identification of some problem requiring attention and action, and the promotion of activities to produce a solution. In basic assumption mentality, however, there is a collusive interdependence between the leader and the led, whereby the leader will be followed only as long as he or she fulfils the basic assumption task of the group. The leader in baD is restricted to providing for members' needs to be cared for. The baF leader must identify an enemy either within or outside the group, and lead the attack or flight. In baP, the leader must foster hope that the future will be better, while preventing actual change taking place. The leader who fails to behave in these ways will be ignored, and eventually the group will turn to an alternative leader. Thus the basic assumption leader is essentially a creation or puppet of the group, who is manipulated to fulfil its wishes and to evade difficult realities.

A leader or manager who is being pulled into basic assumption leadership is likely to experience feelings related to the particular nature of the group's unconscious demands. In baD there is a feeling of heaviness and resistance to change, and a preoccupation with status and hierarchy as the basis for decisions. In baF, the experience is of aggression and suspicion, and a preoccupation with the fine details of rules and procedures. In baP, the preoccupation is with alternative futures; the group may ask the leader to meet with some external authority to find a solution, full of insubstantial hopes for the outcome.

Members of such groups are both happy and unhappy. They are happy in that their roles are simple, and they are relieved of anxiety and responsibility. At the same time, they are unhappy insofar as their skills, individuality and capacity for rational thought are sacrificed, as are the satisfactions that come from working effectively. As a result, the members of such groups tend to feel continually in conflict about staying or leaving, somehow never able to make up their minds which they wish to do for any length of time. Since the group now contains split-off and projected capacities of its members, leaving would be experienced as losing these disowned parts. In work-group mentality, on the other hand, members are able to mobilize their capacity for co-operation and to value the different contributions each can make. They choose to follow a leader in order to achieve the group's task, rather than doing so in an automatic way determined by their personal needs.

THE MULTIDISCIPLINARY TEAM

I wish now to look at the effects of the interplay between work-group mentality and basic assumption mentality functioning in a particular situation – the multidisciplinary team. Such teams are to be found in both public and private sector settings. For example, a health centre may be staffed by several doctors, a team of nurses, social workers, counsellors, a team of midwives and a number of administrative staff. In industry, management teams will consist of individuals from production, marketing, sales, audit, personnel and so on. In universities and schools, teams consist of staff teaching a range of subjects, the heads of different departments, together with administrators and others.

Teams such as these often have difficulty developing a coherent and shared common purpose, since their members come from different trainings with different values, priorities and preoccupations. Often, too, team members are accountable to different superiors, who may not be part of the team (see Chapter 20). This is an important and yet often ignored reality which leads to the illusion that the team is in a position to make certain policy decisions which, in fact, it is not. Considerable time can be wasted on discussions which cannot result in decisions, instead of exploring ways the actual decision-makers can be influenced in the desired direction.

The meetings of such teams typically have a rather vague title such as 'staff meeting' or 'planning meeting'. Their main purpose may well be simply for those present to 'meet' in order to give a sense of artificial togetherness and cohesion as a refuge from the pressures of work. The use of the word 'team' here is somewhat misleading: there may be little actual day-to-day work in common. Indeed, because of a lack of clarity about the primary task (see Chapter 3), confusion, frustration and bad feeling may actually be engendered by such meetings, interfering with work. The real decisions about work practices are often made elsewhere – over coffee, in corridors, in private groups, between meetings but not in them. Furthermore, such decisions as are taken may well not be implemented, because it is rare for anyone in the group to have the authority to ensure they are carried out.

Task-oriented teams have a defined common purpose and a membership determined by the requirements of the task. Thus, in a multidisciplinary team, each member would have a specific contribution to make. Often, the reality is more like a collection of individuals agreeing to be a group when it suits them, while threatening to disband whenever there is serious internal conflict. It is as if participation were a voluntary choice, rather than that there is a task which they must co-operate in order to achieve. The spurious sense of togetherness is used to obscure these problems and as a defence against possible conflicts. Even the conflicts themselves may be used to avoid more fundamental anxieties about the work by preventing commitment to decisions and change.

BASIC ASSUMPTIONS IN DIFFERENT PROFESSIONS

So far I have referred to basic assumptions as defensive or regressive manifestations of group life. However, Bion (1961) also refers to what he termed the *sophisticated use of basic assumption mentality*, an important but lesser-known part of his theory. Here, Bion suggests that a group may utilize the basic assumption mentalities in a sophisticated way, by mobilizing the emotions of one basic assumption in the constructive pursuit of the primary task.

An example of such sophisticated and specialized use of baD can be found in a well-run hospital ward. An atmosphere of efficiency and calm is used to mobilize baD, encouraging patients to give themselves over to the nurses or doctors in a trusting, dependent way. BaF is utilized by an army to keep on the alert, and, when required, to go into battle without disabling consideration for personal safety. In social work, baF supports the task of fighting or fleeing from family, social and environmental conditions or injustices which are harmful to the client. BaP finds a sophisticated use in the therapeutic situation, where the pairing between a staff member and a patient can provide a background sense of hope in order to sustain the setbacks inevitable in any treatment.

In trying to understand some of the difficulties of multidisciplinary work, it is helpful to understand the different sophisticated uses of basic assumption mentality adopted by the various professions or disciplines that make up a team. Fights for supremacy in a multidisciplinary team can then be viewed as the inevitable psychological clash between the sophisticated use of the three basic assumptions. Each carries with it a different set of values and a different set of views about the nature of the problem, its cure, what constitutes progress, and whether this is best achieved by a relationship between professional and client involving dependency, fight-flight or pairing. Furthermore, individuals are drawn to one profession or another partly because of their unconscious pre-disposition or *valency* for one basic assumption rather than another. As a result, they are particularly likely to contribute to the interdisciplinary group processes without questioning them (see Chapter 12).

Put another way, one of the difficulties in making a team out of different professions is that each profession operates through the deliberate harnessing of different sophisticated forms of the basic assumptions in order to further the task. There is consequently conflict when they meet, since the emotional motivations involved in each profession differ. However, conflict need not preclude collaboration on a task, provided there is a process of clarifying shared goals and the means of achieving these. However, difficulties in carrying out the task for which the team is in existence can lead to a breakdown in the sophisticated use of the various basic assumptions, and instead *aberrant* forms of each emerge. Examining these can illuminate some of the frequently encountered workplace tensions in teamwork.

For instance, medical training involves an institutionalized, prolonged dependency of junior doctors on their seniors over many years, from which the medical consultant eventually emerges and then defends his new independence. This can degenerate into an insistence on freedom for its own sake. The doctor may then operate from a counterdependent state of mind, denying the mutual interdependency of teamwork and the actual dependency on the institutional setting of hospital or clinic. This can extend to other professionals, each arguing for their own area of independence, with rivalry and embittered conflict impeding thought and work on establishing shared overall objectives for the team.

By contrast, the training of therapists – whether of psychological, occupational, speech or other varieties – tends to idealize the pairing between therapist and client as the pre-eminent medium for change. Aberrant baP can lead to collusion in supporting this activity, while refusing to examine whether or not it is in fact helping, or how it relates to the team's primary task. Indeed, therapist and client may remain endlessly 'glued' together as if the generation of hope about the future were by itself a cure.

In social work, the sophisticated use of baF in the productive fight against social or family injustices can degenerate into a particular kind of litigious demand that justice be done, and on 'getting our rights'. Responsibility for improvement is felt to rest not at all with the individual, but solely with the community. Projecting responsibility in this way then disables the client and social worker from devising together any effective course of action: it is only others that must change.

These are examples where the capacity for the sophisticated use of basic assumption activity has degenerated, and the professional's action and thought becomes dominated by its aberrant forms. Each then produces a particular group culture. Aberrant baD gives rise to a *culture of subordination* where authority derives entirely from position in the hierarchy, requiring unquestioning obedience. Aberrant baP produces a *culture of collusion*, supporting pairs of members in avoiding truth rather than seeking it. There is attention to the group's mission, but not to the means of achieving it. Aberrant baF results in a *culture of paranoia and aggressive competitiveness*, where the group is preoccupied not only by an external enemy but also by 'the enemy within'. Rules and regulations proliferate to control both the internal and the external 'bad objects'. Here it is the means which are explicit and the ends which are vague.

CONCLUSION

In a group taken over by basic assumption mentality, the formation and continuance of the group becomes an end in itself. Leaders and members of groups dominated by basic assumption activity are likely to lose their ability to think and act effectively: continuance of the group becomes an end in itself, as members become more absorbed with their relationship to the group than with their work task. In this chapter, we have seen how the functioning of teams can be promoted by the sophisticated use of a basic assumption in the service of work, or impeded and distracted by their inappropriate or aberrant use.

An understanding of these phenomena of group life, perhaps best obtained through the kind of group relations training programmes described in Chapter 4, can greatly assist both the members and managers of multidisciplinary teams, committees and other working groups.

Chapter 3

The organization of work
Contributions from open systems theory

Vega Zagier Roberts

A living organism can survive only by exchanging materials with its environment, that is, by being an *open system*. It takes in materials such as food or sunshine or oxygen, and transforms these into what is required for survival, excreting what is not used as waste. This requires certain properties, notably an external *boundary*, a membrane or skin which serves to separate what is inside from what is outside, and across which these exchanges can occur. This boundary must be solid enough to prevent leakage and to protect the organism from disintegrating, but permeable enough to allow the flow of materials in both directions. If the boundary becomes impermeable, the organism becomes a *closed system* and it will die. Furthermore, exchanges with the environment need to be regulated in some way, so that only certain materials enter, and only certain others leave to return to the outer environment.

In complex organisms, there will be a number of such open systems operating simultaneously, each performing its own specialized function. The activities of these different sub-systems need to be co-ordinated so as to serve the needs of the organism as a whole, and complex superordinate systems are evolved to provide this co-ordinating function, which includes prioritizing the activities of certain sub-systems over others in times of crisis. In the human body, for example, blood flow to the brain will be preserved at the expense of blood flow to the limbs when there is an insufficient supply of oxygen.

The work of Kurt Lewin (1947) in applying these ideas to human systems was extended and developed at the Tavistock Institute of Human Relations in the 1950s, notably by Rice and Miller. Their organizational model provides a framework for studying the relationships between the parts and the whole in organizations, and also between the organization and the environment (Miller and Rice 1967).

ORGANIZATIONS AS OPEN SYSTEMS

An organization as an open system can be schematically represented as in Figure 3.1. The box in the centre represents the system of activities required to perform the task of converting the inputs into outputs. Around it there is a boundary

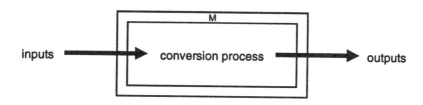

Figure 3.1 Schematic representation of an organization as an open system[1]

separating the inside from the outside, across which the organization's exchanges with the environment take place. These exchanges need to be regulated in such a way that the system can achieve its task, and therefore there needs to be *management of the boundary*, represented by M. For example, an automobile factory takes in or imports raw materials such as steel, and converts them into products or outputs which it exports. This throughput defines its task, namely producing cars.

Obviously, most enterprises are much more complex than shown in Figure 3.1, with many different kinds of inputs and outputs, and a variety of task-systems. The factory, for instance, also takes in information from the environment and uses it to produce financial plans and marketing strategies. It is likely to have different departments such as production, marketing, personnel, and so on, all of which need to be co-ordinated. Furthermore, the tasks of these sub-systems may at times conflict or compete. How to allocate resources and how to prioritize among the organization's different activities is determined by its *primary task*, defined by Rice (1963) as the task it must perform if it is to survive .

This is not as simple as it might first appear. Different groups within the organization may have different definitions of the primary task. For example, the primary task of an automobile factory would seem to be to produce automobiles, and that is how the assembly-line workers are likely to define it. Members of other departments may see it as remaining at the forefront of new technology, or in terms of sales figures. Consequently, the decision to stop producing high-performance cars or to use parts manufactured by another company will be experienced very differently by different members of the organization. If the factory is the principal employer in the local community, its primary task may be regarded by some as providing employment, and the decision to reduce the labour force may well precipitate intractable conflict. Where the primary task is defined too narrowly, or in terms of its members' needs, the survival of the organization can become precarious.

1 The figures in this chapter are adapted from Miller and Rice (1967).

TASK AND ANTI-TASK IN THE HUMAN SERVICES

In institutions which exist to change or help people, the difficulties in defining the primary task precisely and realistically are even greater. They usually have multiple tasks, all of them important and even essential. For example, teaching hospitals must treat patients, train staff and do research; prisons are required to be custodial, punitive and also rehabilitative. There are often conflicting assumptions both inside and outside the organization about which task has priority, and some objectives may even be incompatible. Furthermore, it can be inherently difficult to define aims in other than general terms, such as health, or education, or welfare. This problem is exacerbated by the current dramatic pace of social change and the accompanying changes in the assumptions underlying such terms.

The concept of the primary task can seem to be an oversimplification, given the complexities with which most organizations have to contend. However, it can provide an invaluable starting-point for thinking about what is going on in a group or organization. Miller and Rice described it as 'a heuristic concept', that is, a tool 'which allows us to explore the ordering of multiple activities . . . [and] to construct and compare different organizational models of an enterprise based on different definitions of its primary task' (1967: 62). It is not a matter of saying that an enterprise *should* have a primary task, but rather that at any given time it *has* one which it must perform if it is to survive. As conditions change, the primary task may shift, either temporarily, at times of crisis, or permanently. When changing conditions lead to a permanent change in the primary task, either explicitly or implicitly, this can affect 'the ordering of multiple activities' in ways which have major implications for everyone involved.

Gordon Lawrence (1977) developed this idea of the primary task as a tool for examining organizational behaviour by proposing that people within an enterprise pursue different kinds of primary tasks. The *normative primary task* is the formal or official task, the operationalization of the broad aims of the organization, and is usually defined by the chief stakeholders. The *existential primary task* is the task people within the enterprise believe they are carrying out, the meaning or interpretation they put on their roles and activities. The *phenomenal primary task* is the task that can be inferred from people's behaviour, and of which they may not be consciously aware. Analysis of the primary task in these terms can highlight discrepancies between what an organization or group says it sets out to do and what is actually happening. It can thus serve as a tool for individuals and groups within an enterprise, as well as for consultants, to clarify and understand how the activities, roles and experiences of individuals and sub-systems relate to each other and to the enterprise as a whole.

The confusion in helping institutions and in the society they serve about what their primary task is (or should be) often results in inadequate task definitions, which provide little guidance to staff or managers about what they should be doing, or how to do it, or whether they are doing it effectively (Menzies Lyth

1979; see also Chapter 21). This is a major source of the individual, group and institutional difficulties described in this book. Turquet (1974) warned that when a group does not seek to know its primary task, both by definition and by feasibility, there is likely to emerge either dismemberment of the group, or the emergence of some other primary task unrelated to the one for which it was originally called into being. This *anti-task* is typical of groups under the sway of basic assumptions, as described in Chapter 2, whereas the primary task corresponds to the overt work-oriented purpose of Bion's sophisticated work group. Both are about survival. The primary task relates to survival in relation to the demands of the external environment, while basic assumption activity is driven by the demands of the internal environment and anxieties about psychological survival. In the examples which follow, we will see how lack either of task definition, or of feasibility, or both, can lead to anti-task activity as a defence against anxiety.

Vague task definition

One of the commonest ways of dealing with the problems outlined above is to define organizational aims in very broad and general terms.

The Tappenly Drug Dependency Unit was originally set up as a methadone clinic, staffed by nurses and doctors, to help clients reduce and eventually abstain from abusing heroin. In the 1970s, a psychologist and a social worker joined the team, and the orientation began to shift towards a more holistic approach, with attention to addicts' psychosocial as well as their medical needs. In the 1980s, concerns about HIV infection shifted the emphasis in drug work across the country away from emphasis on abstinence *per se* towards 'harm reduction'. Counsellors were hired to join the multidisciplinary team and long-term support to help clients stabilize their life-style became the predominant form of treatment. The stated aim of the unit became 'to provide a comprehensive service to residents of Tappenly with drug-related problems'.

Over a two-year period, the team almost doubled in size, yet the waiting list continued to grow longer. The staff felt the only solution was to recruit yet more staff, despite evidence that this shortened the waiting list only briefly, until the new workers had full caseloads. Instead of hiring more staff, the managers suggested that the team review and revise their working methods and curtail the amount of counselling offered to clients, which the team experienced as devaluing and undermining their good work. They continued to struggle to provide as much counselling as was needed to as many clients as possible, repressing their anxiety about the people on the waiting list who were getting no support at all. When the hospital management finally intervened actively in the crisis by imposing a limit on the number of counselling sessions any one client could have, team members felt outraged, betrayed and demoralized.

The statement of their aims, which the team had written in such a way that any and all of their activities were covered, was of no help in determining how best to deal with drug problems in Tappenly. The team behaved as if their task were to provide as much help as was needed to everyone who needed it. As the demands on the unit increased in both scope and numbers, this became impossible. In turn, this created a great deal of stress for the staff: both the conscious stress of ever-growing caseloads, and the unconscious stress of failing to help the many addicts on the waiting list. Their only recourse was to blame the shortcomings of the service on the managers' refusal to provide them with sufficient resources.

Defining methods instead of aims

Another common way of dealing with the difficulties in defining the aims of helping organizations is to define their methods rather than what these methods are intended to achieve.

Pathways, a service for young people aged sixteen to twenty-four in a deprived inner-city area, defined its aim as 'to provide free information, advice, and counselling to young people in difficulty, especially those suffering from the results of social deprivation and prejudice'. This service had two parts: social workers gave advice and information about housing and benefit rights, equal opportunities, career training and other community resources; volunteer counsellors and therapists provided support for psychological problems.

There was constant conflict between the two groups, ostensibly about how the premises were used. The social workers ran the advice and information service as a drop-in centre, encouraging clients to stay around to meet each other informally. This made for a good deal of noise and sometimes so much crowding that it was hard to get through the door. The counselling service was based on an appointment system, with clients being encouraged to arrive and leave promptly, and the counsellors complained that the noise interfered with their work, which required a quiet and reassuringly calm atmosphere. The ill feeling could be dealt with only informally, since there were no staff meetings at which all workers met together.

Consultants, called in to help with an organizational review, noticed how each group spoke disparagingly about the other, but in fact were largely ignorant and even misinformed about each other's training, skills and ways of working. The social workers worried that the counsellors encouraged futile navel-gazing which might well undermine the clients' capacity to fight for their rights, while the counsellors considered that the emphasis on resisting social injustice prevented the young people from taking responsibility for their own lives and fostered an expectation of being taken care of. As a result, there was no cross-referral between the two services. New clients whose first contact was to drop in would meet a social worker. If, on the other hand, the first contact were by telephone, they would be more likely to be allocated to a counsellor.

It seemed to the consultants that neither view was accurate and that the two groups were doing potentially complementary work. They therefore worked with the organization as a whole on developing a definition of their primary task towards which both sets of activities would contribute. Meeting together as a single staff group enabled workers to develop more understanding and respect for each other's work, and to focus on the question of what the organization was trying to achieve, rather than on which method of work was superior. This not only improved relations among the staff, but was also of benefit to clients, who began to be cross-referred when appropriate, and thus to have access to the specific help they needed.

By splitting up the task of supporting young people in developing the resources – both internal and external – necessary for taking up an adult role in society, and by defining its methods rather than its aims, Pathways defended itself from anxieties about failing to remedy some of its clients' problems. Instead there was conflict and rivalry among staff which interfered with effective work. Furthermore, the aims as originally stated could do nothing to resolve the conflicting assumptions about the cause of the clients' difficulties.

Avoiding conflict over priorities

Another way of avoiding dealing with disagreements within an organization is to define the task in a way that fails to give priority to one system of activities over another.

Early Days Nursery School defined its task as 'providing quality care and education for under-fives'. Staffed by both nursery nurses and nursery teachers, it had a high ratio of staff to children, and an excellent reputation, particularly for its standard of teaching. When funding was suddenly severely reduced, there was great uncertainty and conflict about how to use the now limited resources. Should the higher-paid teachers be replaced by more nursery nurses? Or should the ratio of staff to children be reduced? Or should they spend less on equipment, or move to smaller premises in an effort to maintain the standard of teaching at all costs? Different people had quite different ideas about priorities, and latent disagreements about what constituted quality care and education, and what resources were required to produce these, erupted, largely as painful personal attacks.

At this point, it seems useful to make a distinction between the terms 'aim' and 'primary task', which are to some extent used interchangeably in this and subsequent chapters. More precisely, aims are broadly outlined statements of the intended direction of an enterprise, while primary task refers to the way in which the system proposes to engage with these aims. Thus, Pathways might say that its aim was to provide opportunities for the development of under-fives. One way of engaging with this would be to facilitate the children's transition from home to primary school. In open systems terms, the inputs in this case would be children under five who were not yet ready for school, and the outputs would be school-ready five-year-olds. This would then open a crucial debate within the nursery about what Pathways regarded as the most important features of school-readiness, which in turn would assist it in making decisions about 'the ordering of multiple activities'.

Failing to relate to a changing environment

When the external environment into which an organization exports its products changes, it may become necessary for it to revise its primary task. Failure to do so will cause stress and compromise the organization's effectiveness.

> Putlake High School had long been considered an outstanding school, with a reputation for innovative programmes, including work experience, personal and career counselling, and an individual tutorial system. When funding cuts and restructuring of the county's educational system threatened these programmes, there was a steep decline in staff morale and an external consultant was invited to work with the teachers on how to cope with the recent changes.
>
> On her first visit, the consultant was struck by the quiet everywhere; there was little contact between teachers and pupils outside the classroom, and there were no telephones in the staff rooms. In fact, it was quite hard to find anyone to speak to, or to reach staff members by telephone. Interviews with individual teachers revealed their great anxiety about what would happen to their pupils when they left, and despair about their efforts to equip the pupils with skills they would not be able to use, given the high local unemployment. The teachers were cynical about their task, seeing it as 'keeping the kids off the streets for an extra year or two' and 'keeping unemployment figures down'.

Here the staff were continuing to work at the now outdated task of preparing their students for the world of work. To cope with their guilt and anxiety, they avoided contact both with the students and with the outside world, clustering together to whisper complaints about their managers. Most had been at the school for over ten years; it was as if the primary task of the school had become to provide the staff with secure jobs in a citadel cut off and protected from the terrifying world outside the school. It had become a closed system, existing primarily to meet the needs of its members: transactions across its boundaries with potential employers, former students and parents were kept to a minimum. Internal boundaries, too, notably those between staff and managers, and between staff and pupils, had become rigid, defending the staff from anxieties and information that they could no longer manage.

TASK AND ANTI-TASK BOUNDARIES

Where there are problems with the definition of the primary task, there are likely also to be problems with boundaries, so that instead of facilitating task performance, they serve defensive functions. In the last example, the boundary around Putlake High School had become relatively impermeable, so that the necessary exchanges with the environment no longer took place. In the next example, boundaries were located and managed in such a way that they failed to relate the parts to each other and to the overall enterprise.

> Cannon Fields, the Community Mental Health Centre (CMHC) for Northwest Wresham, was one of three such centres – the others serving Northeast and South Wresham – set up to provide mental health services in the community as part of a plan to close wards at Wresham Psychiatric Hospital. The team, most of whom had previously worked at the hospital, were very committed to creating a service where clients would flourish, developing social and independent living skills which would give them a better quality of life as well as preventing breakdown and hospitalization. They regarded the hospital as rigid, oppressive and suppressive of individuality, and based their programme planning on the intention to be as different from it as possible. The clients could come and go freely, choosing whether or not to attend formal therapeutic activities. Similarly, staff could work as they chose, with individuals or groups, chronic or acute patients. It was difficult to set any limits or enforce any decisions, lest this curtail individual freedom, evoking the spectre of the 'bad old days' on the wards.
>
> The aim of Cannon Fields was defined as 'offering a comprehensive community-based mental health service to residents of Northwest Wresham with emotional and psychiatric problems, and to prevent admission to hospital'. As a result, admission of anyone in their catchment area to Wresham Hospital was experienced by the staff as a failure, which they tended to blame on bad management, inadequate resources or incompetence on the part of general practitioners, casualty room staff and other professionals.
>
> Relations between the centre and the hospital's C ward, to which Northwest Wresham residents were admitted when they needed in-patient care, were antagonistic. Cannon Fields staff considered attending ward rounds at the hospital a disagreeable chore, and left as soon as the round was over, as if contact might contaminate them with something they had been lucky to escape. They spoke of ward staff with pity, but also with contempt, and the ward staff regarded them as stand-offish and unhelpful.

The wording of the aims statement of Cannon Fields placed a boundary between it and the hospital. The first part defined Cannon Fields in terms of being what the hospital was not, namely community-based, rather than indicating what it might be trying to do. The second part, 'to prevent admission to hospital', on the

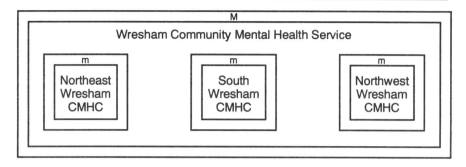

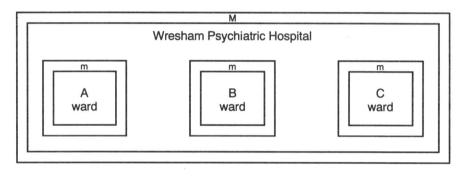

Figure 3.2(a) Organization of Wresham Mental Health Services before restructuring

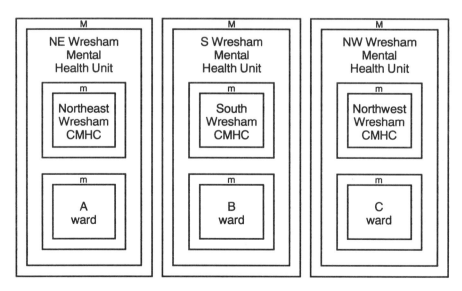

Figure 3.2(b) Organization of Wresham Mental Health Services after restructuring

one hand was too narrow a definition, and on the other hand set an impossible task. Together, the two parts supported the staff ideal of being a superior alternative to hospital care, as different and separate from it as possible. The team behaved *as if* the task it existed to perform were to do away with the need for a hospital altogether.

This as-if task was supported by the management structure. Community psychiatric care was managed as a system quite separate from, and even in competition with, the hospital-based psychiatric service (see Figure 3.2(a)). Subsequently, the management structure was altered to correspond to the three catchment areas (see Figure 3.2(b)). The new boundaries matched and supported the task of providing a comprehensive mental health service, comprising both in-patient and community services, to each catchment area. Patients could then be more readily seen as a shared responsibility, whether they were at any given moment in the hospital or in the community, and the rivalry between the hospital and community lessened. Ward rounds became a central activity for the staff of Cannon Fields as well as of C ward, involving their working together at assessing the needs of their joint clients.

MANAGEMENT AT THE BOUNDARY

The management of boundaries is absolutely crucial to effective organizational functioning. Boundaries need both to separate and to relate what is inside and what is outside. Where an enterprise consists of multiple task-systems, there is a boundary around the system as a whole, as well as one around each of the subsidiary systems, and each of these boundaries needs to be managed so that all the parts function in a co-ordinated way in relation to the overall primary task (Miller and Rice 1967).

Whereas most organizational charts place managers above those they manage, the open systems model locates them *at the boundary* of the systems they manage

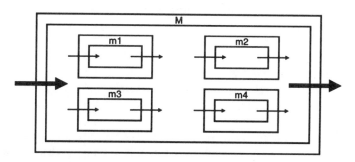

Figure 3.3 Management of multiple systems of activity within an organization

Note: The smaller boxes represent discrete task-systems, each with a boundary managed by m1, m2, etc. The larger box represents the overall enterprise, managed by M. In conventional organizational terms, M is the line manager of the four m's.

(see Figure 3.3). It is only from this position that they can carry out their function of relating what is inside to what is outside the system. This includes being clear about the primary task, attending to the flow of information across the boundary, ensuring that the system has the resources it needs to perform its task, and monitoring that this task continues to relate to the requirements of the wider system and to the external environment.

The manager who loses this boundary position, either by being drawn too far into the system, or by being too cut off, can no longer manage effectively. This is also the case for each individual in an organization. Even those who are not in designated management roles need to manage the boundary between their own inner world – their wishes, needs and resources – and external reality, in order to take up their roles (Grubb Institute 1991).

CONCLUSION

In trying to make sense of the often very confusing processes in institutions, it is useful to start with the question, 'What is the primary task?' This often proves surprisingly difficult to answer, as many of the examples in this chapter have illustrated. However, using the model depicted in Figure 3.1 (p. 29) can help to identify the dominant throughput. In a human service enterprise, this dominant throughput is likely to be people, who enter the system in one state and – as a result of the 'conversion' or 'transformation' process within – leave in a different state. Thus, defining the primary task requires thinking first about what this intended or desired 'different state' is, and then about how the system (group, department or organization) proposes to bring this about.

The next question is, 'How does our way of working relate to this task?' If it does not, it can be helpful to ask 'What are we behaving *as if* we were here to do?' Identifying this as-if task can provide clues about the underlying anxieties, defences and conflicts which have given rise to the dysfunctional task definition and to the associated dysfunctional boundaries. These are questions not only for managers and consultants, but also for the individual worker. To be personally effective in our work roles, we need to be clear about the task we have to do; to be able to mobilize sufficient resources, internal and external, to achieve it; and to have some understanding of how our own task relates both to the task of the system in which we are working and to the task of the institution as a whole. These questions pave the way for asking 'How well are we doing?' (see Chapter 21) and for beginning to think about how we might engage differently with the task in hand.

We have seen many examples in this chapter of groups and organizations whose primary task lacked definition or feasibility or both. When managers fail to maintain their position at the boundary, they are likely to get caught up in unconscious group and institutional processes, and will not be able to ask the questions that need to be asked. They will then be unable to do what needs to be done to restore to the system the capacity for effective work.

Chapter 4

Authority, power and leadership
Contributions from group relations training

Anton Obholzer

It is self-evident that clarity in matters of authority, leadership and organizational structure is essential for the competent functioning of any organization. Yet confusion abounds. Why should this be so? What are the factors contributing to the perpetuation of this state? In this chapter we shall look at these core concepts, their derivation and interrelationships, and the individual and organizational factors that commonly lead to the appearance and persistence of institutional anti-task phenomena. The way these concepts are used throughout this book is grounded in a model of group relations training described at the end of this chapter.

AUTHORITY

Authority refers to the right to make an ultimate decision, and in an organization it refers to the right to make decisions which are binding on others.

Authority from 'above'

Formal authority is a quality that is derived from one's role in a system and is exercised on its behalf. For example, the director of a company derives authority from the board of the company. The board makes the appointment, holds the director responsible for outcomes, and also has the power to sack him or her. The board, in turn, is elected by, and thus derives its authority from, the shareholders of the company. Thus, authority derives from a system of delegation, in this case from the shareholders to the board, and from the board to the director. Usually there is a constitution or other formal system which lays down terms of office and other mechanisms for delegating authority. These are human-made systems, agreed through common consent. If the system becomes outdated and is no longer held by common consent, it has to be changed to take the new factors into account.

However, not all systems are so clear-cut in hierarchical terms. For example, in voluntary organizations there are often a number of 'stakeholders' – funders, management committee members, clients, staff, professional groups, referring agencies and so on – which may all, in different ways, claim ownership of the organization. Some of these will hold different views on where authority ultimately comes from (or should come from), to whom it is delegated, and to what extent.

Authority from 'below'

Members who voluntarily join an organization are, by definition, sanctioning the system. By the act of joining, they are, at least implicitly, delegating some of their personal authority to those in authority, and in so doing confirming the system.

Authority does not, however, derive only from an external structure as outlined above. It also has internal components which may be explicit and conscious, or unconscious and therefore not available to be worked with. These internal components include the nature and extent of the ambivalence affecting the delegation of authority to those in charge. For example, a chief executive might have full authority delegated to him or her by the authorizing body, but there may have been no dialogue or consultation with the working membership of the organization regarding the appointment. In this case, the latter may accept the concept of management and sanction the authority of the role, but not that of the person in role. The withholding of authority from below, in the form of not sanctioning, means that full authority cannot be obtained, and that there is an increased risk of undermining and sabotage.

Of course, 'full authority' is a myth. What is needed might be called 'full-enough' authority, to coin a term derived from Winnicott's (1971) concept of 'good-enough' mothering. This would imply a state of authority in which there would be an ongoing acknowledgement by persons in authority (in their own minds, not necessarily publicly), not only of their authority, but also and equally of the limitations of that authority. An integral part of this state of mind would be an ongoing monitoring of authority-enhancing and of authority-sapping processes in the institution.

> A head teacher learned that an informal meeting had been called at school, after school hours, in which the teachers attending planned to discuss the new curriculum. As this had previously been decided at a policy meeting chaired by him, he now faced some difficult questions. Was this to be interpreted as evidence of staff initiative, of their taking their authority to extend the debate, and thus enhancing the head's aims? Or was it to be interpreted as an undermining of authority and formal decision-making structures?

The head's understanding of the meaning of the proposed meeting would probably determine not only what action he would take, but also its outcome.

Authority from within

Apart from the delegation of authority from above and sanctioning from below, there is the vital issue of the authorization or confirmation of authority from within individuals. This largely depends on the nature of their relationship with the figures in their inner world (see Chapter 1), in particular past authority figures. The attitude of such 'in-the-mind' authority figures is crucial in affecting how, to what extent and with what competence external institutional roles are taken up. For example, an individual might be appointed to a position of authority, sanctioned from below, yet be unable to exercise authority competently on account of an undermining of self-in-role by inner world figures. Such 'barracking' by inner world figures is a key element in the process of self-doubt, and, if constant and evident, is likely to prevent external authorization in the first place.

> An accountant who was perceived by both his peers and seniors as very competent, having gone through all the correct processes, was promoted to director of finance, a move approved by his colleagues. Once that promotion had taken place, his work seemed to falter. He had lost his membership of the office club, and in himself felt he was not really up to the job, and that his former peers were now making snide comments. There was no evidence in reality that this was going on. However, snide remarks were an integral part of his relationship with his father; as a child, he had been at both the giving and receiving end of this process.

The opposite dynamic also exists, with inner world figures playing into a state of psychopathological omnipotence, which makes for an inflated picture of the self as regards being in authority, and is likely to produce authoritarian attitudes and behaviour.

> A doctor, appointed at an early age to a consultant post, became increasingly pompous, arrogant and hard to bear by staff and patients alike. His incapacity to listen, to learn from his own or others' experience, and thus to modify his behaviour, arose from an inner world constellation in which he was mother's only child, her adored companion who could do no wrong. Transferred to outer world behaviour, the consequences were disastrous.

There is an important difference between the terms authoritative and authoritarian. Authoritative is a depressive position state of mind (see Chapter 1) in which the persons managing authority are in touch both with the roots and sanctioning of their authority, and with their limitations. Authoritarian, by contrast, refers to a paranoid-schizoid state of mind, manifested by being cut off from roots of authority and processes of sanction, the whole being fuelled by an omnipotent inner world process. The difference is between being in touch with oneself and one's surroundings, and being out of touch with both, attempting to deal with this unrecognized shortcoming by increased use of power to achieve one's ends.

Good-enough authority, at its best, is a state of mind arising from a continuous mix of authorization from the sponsoring organization or structure, sanctioning from within the organization, and connection with inner world authority figures.

POWER

Power refers to the ability to act upon others or upon organizational structure. Unlike authority, it is an attribute of persons rather than roles, and it can arise from both internal and external sources. Externally, power comes from what the individual controls – such as money, privileges, job references, promotion and the like – and from the sanctions one can impose on others. It also derives from the nature of one's social and political connections: how many individuals of prominence can be summoned to one's aid in role. Internally, power comes from individuals' knowledge and experience, strength of personality, and their state of mind regarding their role: how powerful they feel and how they therefore present themselves to others.

In all these, the perceived power or powerlessness counts more than the actual, both of which depend on the inner world connectedness mentioned previously. For example, powerlessness is often a state of mind related to problems with taking up authority. At times there is an interplay between this state of mind and an actual lack of external resources that could otherwise be used to bolster power. However, an individual in a state of demoralization or depression may well have adequate external resources to effect some change, but feel unable to do so on account of an undermining state of mind. In this case, power is projected, perceived as located outside the self, leaving the individual with a sense of powerlessness. By contrast, someone who attracts projected power is much more likely to take – and to be allowed to take – a leadership role. The nature of the projections will affect whether that person is hated and feared, or loved and admired (Grubb Institute 1991).

Power, authority and language

The terms authority and power are often used interchangeably, leading to confusion. They are different, although related, and in organizations both are necessary. Authority without power leads to a weakened, demoralized management. Power without authority leads to an authoritarian regime. It is the judicious mix and balance of the two that makes for effective on-task management in a well-run organization.

The title given to the person in authority in an organization generally gives some indication of the authority/power ratio. Thus 'dictator' makes it clear that the essential component is power. 'Director', 'manager' or 'chairman' generally imply a mix of authority and power. By contrast, 'co-ordinator', a title often given to the most senior manager in voluntary sector agencies, suggests that the person can only take decisions if everyone agrees to them – an unlikely phenomenon – and that there is very little power and capacity to exert sanctions. The choice of such a title may well express ambivalence in the organization about the amount of authority and power it is prepared to give its office-bearer. In addition, the type and length of appointment can radically affect how the post is perceived in terms of power and authority. For example, there is a considerable difference between being appointed general manager and acting manager, or between being given a fixed term or permanent contract.

Clarity of structure and of the constitution make it possible to assess whether or not the system of authorization is functioning, and what steps would need to be taken to withdraw authorization, should that be decided. This is, of course, not possible in authoritarian regimes, where the constitution either does not exist or else is subverted, and rule or management is on the basis of power rather than of law. Furthermore, there needs to be a match between authority and power, and responsibility. Responsibility for outcomes involves being answerable or accountable to someone, either in the organization or else in one's own mind as part of an inner world value system. A sense of responsibility without having adequate authority and power to achieve outcomes often leads to work-related stress and eventually burn-out.

In assessing the nature and functioning of an organization, whether as a member or as an outside consultant, the time used in clarifying the nature, source and routing of authority, the power available, and the names describing various organizational functions, is time well spent.

LEADERSHIP

Leadership and management are also terms that are often used interchangeably. It is true that they have a 'headship' function in common, but management generally refers to a form of conduct by those in authority that is intended to keep the organization functioning and on-task, while leadership also implies looking to the future, pursuing an ideal or goal. Furthermore, leadership by definition implies followership, while management does so to a much lesser degree.

The story of Judith and Holophernes in the Apocrypha is an extreme example of leadership, with consequent risks to the followership. When Judith cut off the head of Holophernes, the leader of the Assyrians, and displayed it to the Assyrian army, they behaved as if they had all lost their own heads and were then easily routed by the Israelites. If the Assyrian army had had less of a leadership cult and more emphasis on management, Holophernes could have been quickly replaced and the outcome would have been quite different. Similar difficulties can often be observed after the departure of a charismatic leader in present-day institutions. Followers are left in disarray, and at the same time may withhold followership from the person appointed as a replacement, disabling this person from both leading and managing.

Leadership, followership and envy

Task performance requires active participation on the part of the followers as well as of the leader. A passive, accepting, *basic assumption* state of followership (see Chapter 3), such as one might find in a demoralized organization, is quite different from a state of mind of exercising one's own authority to take up the followership role in relation to the task. The latter implies clarity about the organization's task, and about where one's role fits in with others.

In order to manage oneself in role, the fundamental question is 'How can I mobilize my resources and potential to contribute to the task?' This requires recognition of where one's role ends and another person's begins, the scope and limits of one's own authority, and a readiness to sanction that of others. Rivalry, jealousy and envy often interfere with the process of taking up either a leadership or a followership role. Staff rivalry is a ubiquitous phenomenon. In a misguided attempt to avoid fanning rivalry and envy, managers may try to manage from a position of equality, or, more commonly, pseudo-equality, often presented as 'democracy'. The term is used as if everyone has equal authority. The hope is that rivalry, jealousy and envy will thereby be avoided; the reality is the undermining of the manager's authority, capacity to hold an overall perspective and ability to lead.

Although there is a substantial body of work on envy in intra- and interpersonal relationships, there is little written about its manifestation in institutions. Yet it is clear that envy in institutional processes is one of the key destructive phenomena, particularly in relation to figures in authority. Envy results in a destructive attack on the person in authority, with resultant spoiling of the work arising under the aegis of that person's authority. Typically, the envious attack on the leader is led by the member of staff with the highest naturally occurring quantum of rivalry and envy. This person is unconsciously set up, by means of projective identification (see Chapters 1 and 14) to express not only his or her own destructive envy, but also that of other group members. The dynamic for the institution is then one in which leader and attacker are pushed into a deadlocked fight, while the remainder of the staff take on the role of distressed and helpless onlookers.

In professional settings, the envious attack may take the form of a debate about 'general principles', or 'technical issues' or 'technique', and is presented as if it were in the pursuit of progress, if not of ultimate truth. It is often only with time that the envious, attacking, destructive nature of this process is revealed. The beauty of this particular defensive institutional constellation is that it not only gratifies unconscious wishes but also attacks the pursuit of the primary task, and this reduces the amount of pain arising from the work of the organization. Such anti-task phenomena are often presented as the most progressive, anti-hierarchical, anti-authoritarian, anti-sexist, anti-ageist, anti-racist way of going about the work of the organization. At times, these ideological arguments are little more than a rationalization for the defensive processes associated with envy; at other times they are serious on-task comments about an organization in urgent need of reform. It is essential that there be enough thinking space in the organization for these differentiations to be worked on, and for the resulting understanding to be implemented.

On-task leadership

Leadership and management share a boundary-regulating function (see Chapter 4), which requires relating what is inside and what is outside the organization. Like the two-faced Roman god Janus, the leader must always be looking both inwards and outwards, a difficult position which carries the risk of being criticized by people both inside and outside the system for neglecting their interests. Concentrating solely on one or the other is a more comfortable position but it undermines the role of the leader, and thus the strength of the institution's representation in the outer world.

In addition to the boundary-management function, leadership is directly related to the pursuit of the aims and of the primary task of the organization. (The distinction between aim and primary task is discussed on p. 33.) Without the concept of primary task, whether called that or something else (according to the language of the organization), it is not possible either to have a marker against which the direction and functioning of the organization can be monitored, or to effect the necessary adjustments to this course and functioning. Such monitoring and adjustment are essential functions of leadership, and the leader's authority to carry them out derives ultimately from the primary task. It is only through a consistent and clear monitoring of the primary task that it is possible to develop and maintain on-task leadership, to avoid the abuse of power, and to keep at a relative minimum the occurrence and spread of basic assumption activity in the organization (see Chapter 3). This also implies that, as the primary task changes, so leadership and followership roles may need to change. For example, in an operating team the head surgeon is usually the leader; but if the patient stops breathing during an operation, the anaesthetist needs to take over the leadership until the breathing is restored (Turquet 1974).

THE GROUP RELATIONS TRAINING MODEL

Many of the key concepts referred to in this chapter either originate from, or have been developed and refined through, the experiential study of group and organizational processes in group relations conferences. Central among these is the 'Leicester Conference', first run by the Tavistock Institute of Human Relations in conjunction with Leicester University in 1957. Since then, in addition to more than forty Leicester conferences, there have been numerous other group relations training events of varying length and design in the United Kingdom, the United States and many other countries, adapted from the original model and expanding its application (see Appendix on p. 211). The basic conceptual framework of the Leicester model corresponds to that described in the three foregoing chapters, combining open systems theory, Bion's work on groups, and later developments from psychoanalysis (Miller 1990a, 1990b). However, the influence of these conferences on the understanding of groups and organizational behaviour, and on the practice of organizational consultancy, has less to do with theory than with learning from the conference experience itself.

At the core of all group relations training models is the idea of the individual participant learning from here-and-now experience. Conferences are designed to be temporary learning institutions, giving participants the opportunity to learn from their own experience about group and organizational processes, and their own part in these. Events are planned to be educational and not therapeutic, although personal change may well occur as a 'side effect'. Basic to this work is Bion's concept of *valency* – the innate tendency of individuals to relate to groups and to respond to group pressures in their own highly specific way (see Chapter 3). It is important for individuals to know the nature of their own valency, a group and organizational version of the need to know oneself, in order to be prepared for both the resultant personal strengths and weaknesses as manifested in group situations.

Depending on the nature of the design, and the focus of the event, individuals also have the opportunity to study the nature of intragroup processes in groups of different sizes, and to participate in intergroup activities to learn about intergroup processes. In all these events, members can take up a variety of roles and thus learn about the processes of giving and taking authority, working with tasks and roles, bidding for and exercising leadership, and so on.

In moving from one event to another (that is, from one grouping to another with different memberships and tasks), members also have the opportunity to experience their fellow members in a variety of roles, often behaving quite differently according to the roles they are engaged in and the group process in which they are enmeshed. Similarly, they experience the different behaviour of the conference staff in new roles and settings. Thus they can learn about role and task, and how these affect behaviour and feelings. The process of crossing boundaries, as members move from one grouping or event to another, makes available learning which is applicable to organizational settings and to the management of change and of multiple roles.

Consultancy from experienced conference staff is usually available to help members think about what is happening. In most models, there are also times set aside specifically to give participants the opportunity to review their experiences of the various conference events, and to work out how to apply their learning to their 'back-home' situation (Rice 1965). The membership of these 'application' events is usually made up of participants from different organizations and professional backgrounds, and often also from different countries and cultures. This provides opportunities for members to be witnesses to different styles of perceiving and working at problems, whether from leadership or followership perspectives.

Central to the learning process is the repeated discovery of the presence of irrational and unconscious processes that interfere with attempts to manage oneself, the group, task and roles in a conscious and rational way. Such insights, when experienced in the pure culture of a training event, make for powerful learning from experience. The hope is that, as a result of their greater awareness of unconscious processes and their own part in them, members will return to their 'back-home' work-settings better able to exercise their own authority and to manage themselves in role (Miller 1990a).

CONCLUSION

Clarity about the sources from which authority is derived is important, not least so that at a time of crisis further confirmation of authority, possibly with additional powers, can be sought. Similarly, an awareness of the importance of sanction from below can lead to more dialogue between managers and the workforce. Finally, careful monitoring of one's connectedness with one's own inner world authority figures is also important if 'shooting oneself in the foot' is to be avoided or kept to a minimum.

Effective leadership requires not only an authoritative state of mind to monitor the functioning of the organization against the bench-mark of the primary task. A leader also needs the power to initiate and implement changes as required by a change in social or institutional circumstances, or even, in the light of these factors, to change the primary task of the organization. As part of this process, a system of accountability needs to exist, as does a mechanism for the delegation of authority, an in-house network that allows for the flow of both authority and feedback. By such means, it becomes possible to delegate aspects of the primary task to individuals or teams within the structure, and to call them to account for the nature of their functioning in relation to the overall task of the organization.

Part II

Working with people in pain

INTRODUCTION

A major source of stress for staff working in the helping professions is their constant proximity to people in great pain, whether physical, emotional or both. This stress may be related to conscious anxieties, for example about a client at special risk, or about a mistake one has made. Excessive workloads are another conscious source of stress, leaving staff concerned about the quality of the help they are offering. But other anxieties are unconscious, kept out of awareness not only by personal defences but also by collective ones. These anxieties are stirred up by the nature of the work itself, and the defences to which they give rise can exacerbate stress rather than alleviate it.

Chapter 5 describes how clients rid themselves of their painful feelings, and also communicate aspects of their experience which they cannot put into words, by projecting them into the staff. This can have a profound effect on a staff group. They too can become distressed and deal with this by projection. The whole organization can then become caught up in the same state of mind as the clients it exists to serve. If, on the other hand, the unconscious processes that affect us at an individual and organizational level can be understood, they can be dealt with in ways that further rather than hinder development.

Each of the next six chapters focuses on work with a different client group, and illustrates some of the ways staff relationships and working practices are structured so as to defend against the anxieties inherent in the task. Each chapter also develops themes which readers working with other kinds of clients in other sorts of settings are likely to recognize and find relevant to their own experiences. By describing both the difficulties for which consultation was requested, and their actual experiences and thinking while consulting to these institutions, the authors hope to offer readers new ways of reflecting on their own painful experiences at work.

Chapter 12 shifts in emphasis from considering how the nature of the work affects workers, to exploring what individuals bring to the work – their unconscious needs and the unresolved issues from their past which can make them especially vulnerable to the processes described in the preceding chapters.

Why have we chosen the particular kind of work we do? Thinking about this, individually and with colleagues, can be an important step towards managing our anxieties and stress differently.

Chapter 5

The dangers of contagion
Projective identification processes in institutions

Deirdre Moylan

It is sometimes impossible for people in distress to put into words those aspects of their experience for which they want help. They do not know much consciously about what is troubling them; they only know they are suffering. They may turn up at the therapist's consulting room saying, 'I feel bad. I get depressed. Something is wrong with me.' Institutions sometimes seek consultation in the same way, saying what can be translated as, 'There's something wrong here, but we don't know what it is. It doesn't feel good to work here anymore.' Attempts made by the consultant to clarify the nature of the problem may be strenuously resisted; the wish, as is often the case with individual clients in therapy, is to have the problems eliminated, not clarified. It is evident that the individual or the institution is suffering, but from what? More information is needed.

To discover what is wrong, one needs to listen carefully to their story. It is not just the content of what is presented that gives information, but also the way it is presented, and the mood as the information is conveyed.

> A young woman, Joanna, told me a story of a traumatic childhood event. On one occasion, the atmosphere was one of horror, dismay and disgust. I felt I was with a small, frightened little girl who did not know what was going on. On another occasion, the story was told again and the emotional tone was quite different: there was an air of triumph, contempt for the alleged abuser. This time I felt I was sitting in the presence not of a little girl, but of a judge. On a third occasion, the details of the story were quite exciting, and I felt privileged to be the one entrusted with this information.

The words are the same, the story is the same: what is different is the experience. The different ways in which the story is told offer glimpses of the different levels of meaning that this particular event has for Joanna. These are conveyed not by the words she uses, which may be identical each time, but by creating an experience. So, on the first occasion I felt horrified and desperately sorry for her suffering. The second time, I felt little sympathy, and even somehow guilty, as if I were the one who had committed the offence. I felt dirty and uncomfortable, and wanted to get away. By contrast, on the third occasion I felt rather special, and wanted to hear more. Three very different experiences for the listener, which reflect accurately three different aspects of the experience for Joanna, each requiring understanding. If only the frightened girl gets attention, the story may have to be told again and again until the guilty, uncomfortable girl who feels dirty, or the excited little girl who feels special, can be understood. To start with, however, Joanna was not able to put these difficult and complex feelings into words. Instead, she conveyed exactly what it was like to be there by re-creating the experience. Between us, we re-lived it in the session, painfully piecing it together, and putting it into words that could be thought about, so that the whole experience, in all its complexity for Joanna, could be understood.

The ability to recreate experiences in this way is an important method of communication. It is by no means limited to the psychotherapeutic situation, but is a form of communication we all use, part of our human experience. For small babies, for example, it is usually the only means at their disposal for communicating their needs. Adults tend to get agitated when a baby is crying; they know something of the distress of the baby because they are experiencing it too. This may lead them to want the baby to be quiet, or to try to alleviate the distress. In psychoanalytic terms, this communicative process is called *projective identification* (see Chapter 1): the baby projects the feelings it cannot manage into the mother, so that – through feeling them herself – she can process them on the baby's behalf.

LEARNING TO LISTEN

However, merely sending a message is not enough. Communication also requires that there is someone who is able to hear and correctly understand the message. Too often, we are only able to hear what we expect to hear. Alternatively, we hear what we are comfortable with, and screen out the rest. In listening to a story like Joanna's, for example, we may be able to hear only the distress of the molested child, and not the communications about excitement or triumph, which we find more disturbing. The painful story is therefore not fully understood by either, and so gets repeated endlessly. What needs attention is the listener's *own* experiences, or *countertransference* (see Chapter 1), as the story is told. This conveys the essence of the trauma, how painful it was to be there, and can make it possible to discover the exact nature of the pain. The capacity to hear the message accurately requires the ability to pay attention to all aspects of one's experience, and depends on many things. For example, the mother of the crying baby may or may not be able to respond accurately to the communication, hearing the difference between a cry for food, and one of fear or loneliness. If she is tired or preoccupied, her capacity to hear – and therefore her ability to respond appropriately and to 'contain' the baby's distress – may be impaired.

We can see similar processes in operation in many other situations. Thoughts and feelings are often stirred up by our contacts with others. Sometimes these are ignored, perhaps because we find them too disturbing. At other times we pay attention, and they can greatly influence our subsequent behaviour. For example, while listening to a charming sales representative, a potential buyer has a brief image of the salesman back at his office, laughing with his colleagues. The image might alert the buyer to recheck the credentials of both salesman and product, in case the laughter is at the gullible customer. On the other hand, if the idea of oneself as the victim of a confidence trickster is too disturbing, the troublesome image may be ignored.

A fundamental part of the training of psychoanalytic psychotherapists is to learn to attend to the material that patients bring on a number of different levels, including attending to the feelings it evokes in themselves. This is just as important in working with groups and institutions. We can 'hear' and learn a great deal if we are able to attend to atmosphere and to our own feelings, and not just to what is actually being said.

> I was told that the staff of the Daniel Finch Drug Dependency Clinic were finding it a difficult place to work, and I was asked if I would offer consultation. I arranged a first appointment to meet with the staff to find out more specifically what it was that they wanted help with. My most powerful memory of this first encounter is of my horror because it felt like there were dozens of people in the room, all looking at, and to, me to do something about the problems they were having. What could I do for these hordes of people? The door opened and a few more came in – and then again. How many more of them were there? I sat frozen in my chair. I did a surreptitious head count. Sixteen. I had two reactions simultaneously. One was, 'Oh! Is that all? Sixteen is a manageable number.' The other was, 'How can I alone work with sixteen people?' But I remembered I was not quite alone: I worked in a clinic which had a wealth of experience. In this way, I reminded myself that I had a context, which reminded me of my role. Also, I reminded myself that I had worked with larger groups than this in the past, so that my feeling of being overwhelmed by numbers was probably telling me something important about the staff's experience. This enabled me to proceed with the task I was there to undertake, which was to find out what this group of people were looking for when they asked for a consultant.

It takes much longer to describe these first moments than to experience them, but they are worth close attention. They provided an enormous amount of information about the unit, which had been conveyed with great efficiency in the first seconds of the consultation. The atmosphere of a horde conveyed graphically one aspect of what it was like to work within this clinic: my experience mirrored the staff's experience of feeling overwhelmed by the huge numbers of people waiting for treatment. They were trying to deal with a large and demanding clientele at a time when staff quotas were being decreased and demands increased; patient numbers were multiplying and pressure was growing because of AIDS- and HIV-related problems.

Actual figures emerged much later to confirm this first impression of being swamped with clients. Over the previous year, the team had seen about thirty new referrals every month. The average weekly caseload for full-time members of staff was twenty-six. The reality of this in terms of the amount of pressure on each worker, the number of case histories to remember, the level of pain witnessed, the projections and countertransference to be managed, all added up to a monumental task. Indeed, the sense of being overwhelmed emerged as one of the chief stresses of working in this clinic.

One might speculate that the opening moments of a first encounter with a new group will always contain some of this sort of anxiety. I will contrast it, therefore, with another beginning.

> Some nurses had requested a group in which to think about their difficulties working on Sheehan Ward, part of a prestigious and specialized unit offering a particular form of cancer treatment. When I arrived on the ward for the first meeting, arrangements for which had been made well in advance, nobody seemed to be expecting me. When I introduced myself there was puzzlement, followed by recognition. I was then shown to a small room and was told I would be joined shortly by the group of nurses involved. There was a television blaring loudly, and a blackboard bore a large notice saying, 'Support group with Dedrie Maybler today at 2.30.' Only the time was correct, but even this clearly had little impact since no one had arrived. I felt quite uncomfortable: no one seemed to know who I was or what I was there for, despite the preliminary negotiations. Eventually, some nurses drifted in, talking among themselves. They made coffee, ignoring both me and the loud television, as if I had become invisible. I began to feel angry and impotent – I did not know what to say. What I experienced in a dramatic and painful way in these opening minutes later turned out to reflect one of the major concerns of this group. It was not until several sessions later, after the following story had been told, that I was able to understand it.

> One of the nurses, Mary, spoke very movingly about how pained and guilty she felt going into the room of Ulrich, a patient who did not speak English. Such patients were not unusual on this ward. She described how awful she felt undertaking a treatment procedure while ignoring the young man's obvious distress and fear because she was not able to talk to him about it. She quickly did her work, left the room, and avoided Ulrich if she possibly could, although she was aware he was frightened and lonely, and did not know what was happening to him. Her experience was of helplessness because she was unable to use her usual skills to inform or reassure the patient. Like Ulrich, Mary was unable to communicate; like him she felt frightened and impotent, and sometimes angry that she had to tolerate this painful experience she had not expected. She also felt guilty that she was not meeting her own high standards of care and concern.

It seemed that Mary's experience must have been very close to Ulrich's, who – in the absence of a common language – could communicate only with the primitive, unconscious language of projective identification. Ulrich projected his feelings into Mary, who then had to deal with being filled with powerful and painful feelings of impotence, fear, anger and guilt after each contact with him. Mary's wish to do her job to the best of her ability made her particularly likely to have this experience, because she had a strong wish to communicate with Ulrich. In fact, Mary was doing a reasonable job in the circumstances, and the quality of the medical aspects of her work was not diminished. Being helped to understand the unconscious communication going on between her and her patient enabled Mary to feel less overwhelmed by strong emotion when with him, and to understand that she was experiencing what it was like to be a patient on this ward, the feelings of anger and helplessness about having cancer, and the strong desire to run away from it. With this kind of understanding, it can become more bearable to face the situations which stir up the painful feelings, instead of having to avoid them.

Looking back at the beginning of this consultation, we can see how my own experiences mirrored some of the painful aspects of being a patient or a staff member on this ward – in particular, my sense of isolation and impotence, my inability to speak, and my feeling that nobody knew who I was or wanted to communicate with me. Although the difficulties were emphasized with patients who spoke little English, they were also present when all apparently spoke the same tongue. After all, what explanation in any language is adequate for illness, pain or premature death, or when a fifteen-year-old girl's hair has fallen out following chemotherapy? The staff of the ward inevitably experienced helplessness in the face of fatal disease, and difficulty in communicating bad news.

It is therefore not surprising that my initial experience of these feelings was so intense. It was by remembering my own experience that I could understand Mary's impulse to leave Ulrich quickly and have no further contact with him. By staying with instead of avoiding their uncomfortable feelings, it gradually became possible for the staff to understand what they were about, and to tolerate better the reality of working on a ward with cancer patients, where sometimes there is the pleasure of a successful cure, and sometimes the distressing helplessness of watching a patient die.

PROJECTIVE IDENTIFICATION AND INSTITUTIONAL STRESS

These examples have been given in order to illustrate two points. The first is the helpful nature of projective identification, if it can be understood as a communication. One task of the consultant to an institution is to help staff learn to understand and interpret this communication. The second point is to emphasize the difficulties with which the staff have to cope, not least because they are constantly barraged by clients' projections. They need to have adequate and helpful defences of their own; otherwise they are likely to succumb to despair, illness or withdrawal, and to get entangled with the clients in projective identification processes that are not understood and therefore cannot be worked with. The more distressed the client group, the more these unconscious communications are likely to predominate. When there is a lot of pain involved, a natural reaction is to attempt to avoid it, as we saw with Mary, or with the example of the crying baby. Staff groups will tend to avoid understanding or dealing with what is projected into them in this way, and deal with their unprocessed emotions by themselves relying on projective identification as a means of getting rid of what feels too painful. When this predominates, it becomes very difficult for the group to find other ways of coping; it is almost impossible to think clearly, to locate the source of problems, and to find appropriate and creative solutions. In this situation, staff burn-out is also much more likely to become a problem.

I was asked to provide consultation for the Daniel Finch Drug Dependency Clinic at a time when the staff had become demoralized. There was a threat of redundancies, but many were also leaving because of a prevailing sense of hopelessness about the unit. It was difficult for them to find adequate and helpful defences in the service of undertaking their demanding job. Why was their work so difficult? One reason was that, as an everyday aspect of their job, the staff were experiencing and processing painful countertransference feelings similar to those I described occurring in myself at the beginnings of their consultations – but often without being consciously aware that this was what was going on. Very distressed clients constantly project painful experiences into the staff. Without understanding why they felt so hopeless and demoralized, the only recourse for the staff was to leave the clinic, or to attempt to get rid of the pain by avoiding knowledge of it in themselves. A few examples will follow.

Drug addicts frequently live with an internal world full of chaos and uncertainty. Drugs are often used to escape from the experience of this terrible turmoil. Reality becomes distorted, while they convince themselves that, for example, the drug has a beneficial effect on their lives, that it saves them from loneliness, despair and so on. The reality of the damage that the drug does along with the damaging life-style needed to maintain the addiction cannot be tolerated for long. Knowledge of the internal and external chaos is defended against by an assault on truth and reality, which in turn adds to the internal chaos. Those who work with drug addicts are also subjected to this assault on truth and reality; they have to make professional decisions, while living with uncertainty about what is really going on. For instance, they constantly have to make decisions about repeating a patient's prescription for methadone, a drug used to replace heroin while the patient builds up a healthier life-style away from the criminal world of illegal drug acquisition.

Jonathan, a recent referral to the Daniel Finch Clinic, told his key-worker Paul that he needed an extra methadone prescription because he had been mugged and his prescription stolen. Was this the whole truth, or a distorted version of the truth? Was Jonathan mugged or did he turn a blind eye while his girlfriend pocketed the drug? Or had he already filled the prescription and sold the methadone on the black market? If Paul gave the new prescription, would he just have been conned? If he did not give it, was he heaping further injury on someone who was already suffering? Would it then be his fault if, as he threatened, Jonathan committed a crime to obtain money to buy the heroin he needed, since he no longer had his methadone? By now Paul was experiencing doubt, uncertainty, guilt, anger and internal chaos; in fact, his state of mind for the moment mirrored that of the client.

To remain aware of this state of mind, to try to understand it and function professionally at the same time, is very difficult. Small wonder, then, that the staff need to develop defences in order to cope. Some defences are necessary and can serve development, creativity and growth. However, there is a strong pull in these circumstances to use the same defences as the clients. Caught up in projective identification with the clients, the workers can, and do, find themselves operating in similar ways.

One member of the group, Harry, spent considerable time in one of our meetings talking about plans for the future of the unit, and projects he would undertake in the coming year. Others joined in the discussion, which was lively and stimulating. Meanwhile, I found myself feeling confused. I thought Harry had given in his notice, and would be leaving the unit shortly. I was full of doubt. Was it my imagination that he was leaving, and if so, why would I imagine such a thing? I was tempted to say nothing and let the discussion run on, but something felt wrong and I decided to express my doubts.

Notice the similarity of my state of mind to the state of uncertainty and confusion described earlier. Like the staff, I was asking myself if I could trust my own memory, my own instinct that something was wrong – just as they had to ask themselves constantly if they could trust their own memories and their instincts. The state of mind of the patient was mirrored in the staff, which was then mirrored in the consultant.

I asked Harry if he was leaving the unit shortly. He looked surprised, and agreed he was – in fact, within a couple of weeks. He then thought about the way he and the others had been talking. He had not forgotten he was leaving, but said it felt like it would not happen for a long time; it was as if he had no sense of the reality of time. Then he remembered one of his long-term clients, Ian. When he heard Harry was leaving, Ian was very upset. Eventually he was able to ask when it would happen, and was told it would be in three months' time. 'Three months!' said Ian, 'But that's forever!' And the sense of upset disappeared. Ian had completely obliterated the reality that time would pass, that three months would go by, and that Harry would then leave.

In the staff group, Harry and the group were living out the same defence, planning work as if Harry were never going to leave. All of them were caught up in a projective identification with the client, using the same defence of denial of reality. By understanding what was happening, we were able to see that this denial served to protect the group for the moment from the pain of the impending separation and loss. The reality, however, was that the separation would take place, and the team needed to prepare themselves for the loss of a highly valued colleague. Harry, too, needed to prepare himself for the loss of a job and colleagues he had enjoyed. Only when reality was faced could the future be considered; for example, the question of how to recruit a suitable replacement for Harry could now be thought about. By being more in touch with his own reluctance to deal with the painful realities of leaving, Harry was also able to understand better his client's fear that to face his own pain about the loss of his key-worker would be intolerable. Work could now begin with Ian too, to prepare him for what would happen, and to plan more realistically for his future therapy.

Here we see how projective identification can be used as a defence against facing a reality that feels unbearable. The same defence was being used by the patient and the staff, and, for a while, by me when I was tempted to avoid the problem also. In situations like this, it is as if the whole system gets caught up in something contagious, a state of mind that is passed rapidly from one person to another until everyone is afflicted and no one any longer retains the capacity to face the pain of reality.

One of the tasks of the consultant is to use his or her own feelings to understand the staff's experiences, and to help them recognize how they become caught up in projective identification with their clients. By noticing and reflecting on their experiences, rather than avoiding them, the staff can free space for thinking about task-appropriate ways of going about their work.

For instance, when I first began consulting to the Daniel Finch Clinic, the staff tended to blame all their difficulties on incompetent management, upon whom they seemed angrily dependent. They were not able to deal directly with the managers about whom there was so much complaint, or to take decisions for themselves, projecting all the feelings of helplessness and incompetence arising from the work itself into the management group outside the unit. When they began to recognize this, they could start to separate their own emotional responses from those of their drug-dependent clients, and their inappropriate dependency on management lessened considerably. Complaints or questions were taken up actively with managers, rather than being left as a source of helplessness and resentment. Workers also took on more self-management, negotiating among themselves and with other units in the wider organization to improve their service. Being less caught up in projective identification with their clients, they were able to function in a more creative, satisfying and efficient way.

CONCLUSION

This kind of understanding is not limited to consultants. Managers or other members of a team can learn to stand back from their experiences and to use their feelings to understand what is going on. By knowing about ways in which the institution can become 'infected' by the difficulties and defences of their particular client group, staff are more likely to be aware when this is happening, and to use their feelings to tackle their problems in a direct and appropriate way, rather than resorting to avoidance or despair.

Although the examples in this chapter have been of organizations both of which deal with very ill patients, the processes described are evident in all organizations, whatever their task and whatever their client group. The other chapters in this part describe other settings, each with its own particular unconscious pressures to which staff are subjected, and its own unconscious organizational defence systems. Awareness of these opens up the possibility of choice. Instead of denial and projection, there is room for thoughtful and creative interest in the problems of the institution, and for developing conscious strategies that support healthy growth and development.

Chapter 6

Attending to emotional issues on a special care baby unit

Nancy Cohn

As a child psychotherapist covering a large area outside London, I came to be involved with a special care baby unit following a conversation with the consultant paediatrician responsible for the ward. She had told me she felt the staff were under particular stress because of the nature of their work, and wondered if I could do anything to help. Apart from wanting to be of help to the staff, I had my own reasons for wanting to work on the unit. I had observed that a number of children referred to the child guidance clinic in which I worked had begun life on such units, and I wanted to understand what those early experiences were like.

OVERCOMING SUSPICION

The consultant paediatrician had suggested I meet with the director of nursing services, Ms Larkin, to discuss the idea. Following this meeting, I was sent to see the senior nursing officer for paediatrics, who agreed there was a great deal of stress on the ward. She took me there and introduced me to Wendy, the nursing sister in charge, explaining briefly that I might be of some help, and she then left. My introduction to the ward felt somewhat awkward and abrupt. Might not Wendy and her staff regard my arrival as an unwelcome invasion? My presence had been initiated by managers without consultation with the people I was being sent to help. This probably contributed to the way in which I was originally seen on the ward and the degree of suspicion I had to deal with.

Initially, I went weekly to the ward on a regular day and at a regular time. Each time it was necessary for me to introduce myself all over again and explain why I was there. Because of shift systems, I was always unsure whom I would meet – and in the beginning I often faced an almost entirely new group of staff. Each time I would say to them that it had been thought that working on this type of ward was very stressful for staff and that it might be helpful for them to be able to talk about it. Some staff would say yes, they were under a tremendous amount of stress, but it was clear they felt uneasy about my presence. Other staff reacted in an openly hostile and persecuted way: when I walked on to the ward, they would turn away, as if they were hoping I would not approach them. They would try to look busy or talk together in a way that indicated I was not welcome. They seemed to fear that I was in some way finding them wanting, or even reporting back to managers, perhaps because I was part of the same organization and employed by the same authority. Another factor could have been the feeling among nurses, and perhaps all medical staff, that they should be able to cope with their emotions, and that discussing (or even having) them meant they were not coping. In that sense, my presence felt like criticism.

I found that my not being a nurse and not being able to enter into the work of the ward, or even to know in detail what the nurses did, was useful. It enabled me to ask basic questions, which sometimes led on to their thinking about aspects of the work which had formerly not been noticed. For instance, one day I saw a nurse massaging a baby's chest with an adapted electric toothbrush to clear the congestion in his lungs. In response to my asking if the baby liked it, the nurse became more aware that the baby was indeed enjoying this, and she then seemed to enjoy the procedure more herself. The nurses had many thoughts and feelings about what they were doing with the babies, to the babies and for the babies; my interest seemed to offer them a chance to explore these.

At this stage, I did not feel my presence was particularly welcomed, but it did not seem inhibiting either. Our talks continued in this casual way for some time. Either we sat around the desk in the nursing station, or I would walk around and approach nurses as they were working, asking them how things were or what they were doing. I spent a lot of time observing them at work, trying to keep out of the way while being close enough for anyone who wanted to talk to do so easily.

Then there was a turning-point. I had mentioned to Wendy that I would be interested in meeting the consultant paediatricians, and after about eight weeks she said she would be introducing me to Dr Miller when he came on his ward round that day. As the doctor and his entourage came past, Wendy stopped them and introduced me as 'the child psychotherapist coming to the ward'. Dr Miller looked somewhat surprised and wondered whether the babies were not a bit young for me to be working with. Wendy immediately replied, 'Oh, she's not here for *them*; she's here for *us* – because of the stress of our job.' This seemed to mark a degree of acceptance of me by Wendy, and, in turn, by her staff.

DOING SOMETHING BY DOING NOTHING

In contrast to the constant activity on the ward, I was sure it sometimes seemed unhelpful to the nurses for me to be standing around when what they needed was another pair of hands. I often felt tempted to find something practical to do, and thought this was very likely how they felt – that looking after the emotional and psychological needs of the parents and babies might not seem like 'real' work to them either. They appeared to feel guilty if they were just being with the babies or mothers, rather than 'doing something'. There were so many urgent, practical, necessary procedures that needed to be followed that it was easy to see how emotional needs could be regarded as almost irrelevant. The primary task of the ward, after all, was the infants' survival. It seemed there were many nurses who would have liked to sit and talk with the parents more, hold the babies, hold the parents by means of their words, but these instincts were often overridden. Yet when they were not able to have contact with the families on an emotional as well as a physical level, the nurses could become mechanical, and sometimes appeared hard.

One example of this is the way they dealt with babies dying. Very ill babies were often sent to London, so deaths on the ward were not so frequent as they were elsewhere; however, they did occur regularly. One day when I came on the ward, I was told about an infant who had recently died. One of the nurses said it had been a particularly distressing experience: she had wanted to make the death as bearable for the parents as she possibly could, and felt she had failed to do so. I asked what happened when a baby died on the ward, and a discussion followed about the procedure. The baby would first be put into the sluice room, and then taken down to the mortuary where the parents could see their child. In the mortuary, there was a cot for children but it was far too large for babies.

It turned out that none of the staff individually were happy with this procedure, but they had never discussed it before. They told me they would have liked the parents to be able to spend some time with their baby on the ward before the body was moved. After our discussion, they decided to get a Moses basket. The parents could then sit in one of the side rooms with the baby, spending as much time as they needed. When I came back to the ward two weeks later, they proudly showed me the basket, which they had lined with pretty bedding. They felt they would be able to handle the next death in a different way, which in fact they did. The nurses could now think much more about making a place for grief on the ward, rather than getting rid of it.

Some time after this, a few of the nurses approached me to say they wanted to offer counselling to bereaved parents whom they had been involved with on the ward, as very little was done for the parents after a baby died. With agreement from Wendy, we arranged to have a series of sessions on bereavement using discussion and role-play. As a result, a new system was set up for supporting parents after the death of babies who had been on the unit. In addition to sending condolence cards and attending the babies' funerals, as they had already been doing, staff now also wrote letters to the parents three and six weeks after the death of their baby, offering visits and counselling, and telling them about a memorial book held in the hospital chapel. I made myself available for discussions with the staff following their visits to the parents.

INDIVIDUAL STRESS

As I became familiar to them, staff began to approach me more easily to talk about particular issues and difficult feelings. One day when I arrived on the ward, Sally came up to me and said she had been lying awake at two o'clock in the morning thinking, 'Thank goodness Nancy will be coming tomorrow.' This was a real surprise for me, as I had not realized that I was becoming important to them, and the extent to which I was now seen as someone who might be able to help them to think about the difficulties they were having.

Sally was under tremendous strain and feeling very anxious about her performance on the ward. She was in a highly charged emotional state, partly because she was undergoing treatment for infertility, and she was often tearful. She told me she was starting to make mistakes and was very worried about the possible consequences. After our discussion, Sally talked to her nursing officer and eventually moved to another ward, having decided not to work with small babies at that time. She also took a few weeks off work and organized some counselling through her GP.

Many of her colleagues had been encouraging her to stay on the ward, feeling it would be good for her to be around babies. They were unaware, as she was, of her anger, hostility and envy towards the mothers who had just had babies, and even towards the babies themselves. However, she was enough in touch with this part of herself to worry that it was affecting her work on the ward, and to realize it could put the babies she was looking after in some danger. After talking with me, Sally felt less as if she were giving up on the babies, and more as if she were looking after them, as well as herself, by having a brief break from work and then going to another kind of ward where the nature of the work would be less disturbing to her. Sally has since told me how valuable the counselling was in helping her to come to terms with some very painful issues which she had been unaware were troubling her. Following this, she conceived without medical intervention.

INTERPERSONAL STRESS

A major area of concern over the years had been to do with staff relationships, both among the nurses and between them and the doctors. When staff feel hopeless or helpless in their attempts to help a baby, they often experience a tremendous amount of anger and frustration. Because this is often unrecognized and denied, and because staff need to protect their patients, their negative feelings erupt instead in relation to each other. There is also a conflict inherent in the kind of work they do which often goes unrecognized, namely reconciling having to do a painful procedure on a baby with wanting to be a caring and kind individual. Having to inflict pain on a baby, the nurses sometimes unconsciously blamed the doctors who prescribed the treatment but did not have to carry it out. Thus the frustration, anger and pain of the work tended to get displaced into conflicts over other issues.

As I became more accepted I was asked to help with some of these conflicts. For example, for a period of time all the anger seemed to be focused on Wendy. Things reached a point where complaints were flying in every direction, and Wendy was feeling persecuted and anxious. At this point, she got in touch with me, explained the situation and asked if I could help. She wanted to understand what she might be doing to provoke all this bad feeling, and I agreed to help her begin to think about this. I explained that I thought the problem probably had at least partly to do with the nature of the role she was in, and arranged to meet individually with her for a number of sessions to explore what might be going on. Later that same week, the director of nursing services telephoned me to ask if I could do something about the difficulties on the special baby care unit. She told me the atmosphere on the ward was very negative, with a lot of conflict and complaints. She was surprised to hear I had been approached directly by Wendy, and that we had already begun to do some work on this issue.

As a result of the discussions with Wendy, I felt that her staff were projecting into her some of the unbearable feelings stirred up by their work with suffering babies and sometimes disappointed or angry parents, especially feelings of guilt and inadequacy. I therefore suggested that we meet together with the senior staff nurses immediately under her in the ward hierarchy to explore these issues and to share the responsibility. It became possible for everyone to see how feelings that seemed to be directed at one or another individual might be to do with that person, but often had more to do with other issues: the work itself, frustration with the health authority, difficult relationships with their own parents or authority figures, and so on. Once the senior staff were more cohesive, we decided to meet with all the staff to discuss the difficulties they had been having with each other and with Wendy. Both Wendy and I now felt confident there would be some support for her from the senior staff, who had previously been allying themselves with the junior staff. They had more understanding of the processes going on within the ward, and could now help to move the group towards thoughtful discussion, rather than blaming and complaining.

These meetings were initially used as a chance to air feelings. It became clear that having discussions about people behind their backs was not helpful to the smooth running of the ward. Many staff talked about going home in a bad mood, and they agreed to try to speak more directly with each other about feelings as they happened, or as soon after as possible, rather than hanging on to these feelings and risking taking them out on the patients, their families or each other.

We also talked about their difficulties with the doctors, who did not attend these meetings or otherwise use me. This was unfortunate, particularly for the junior doctors, who were under tremendous pressure when they came on the ward. They usually had little or no experience of working with small babies, and yet were expected to know what to do and to get on with it. This whole situation was fraught with anxiety for them and for the nurses – and also, of course, for the babies. The nurses talked about having to watch inexperienced doctors trying to do complicated procedures, so that these sometimes lasted for hours instead of minutes. The structure of the relationship between doctors and nurses did not allow the far more experienced nurses to advise doctors on the best ways to do a particular procedure, leading to tremendous stress for all concerned, not least the babies who were 'worked on' for an inordinate length of time. As I could find no way to work directly with the doctors, all I could do was to try to help the nurses contain their anger and distress, and to think about ways they could support the junior doctors more effectively.

MAKING LINKS

Because of my work in the child guidance clinic, I was particularly interested in understanding what problems might occur in bonding when mothers and babies are not able to be together immediately following the birth. There might be things which could be done to encourage better relationships between the mothers and their babies. A referral came to me which enabled me to learn in a vivid way about the experience of one mother and her baby on the ward.

Mrs Pearce was referred by her health visitor, who was extremely concerned about what seemed to be a lack of bonding between Mrs Pearce and her two-month-old daughter, Amy. Mrs Pearce had admitted she was hitting Amy regularly, and there was real concern for the baby's safety. Mrs Pearce, too, was desperate for help. Mother and child had been on the special care baby unit, although I had not met them or heard anything about them at the time. I found it worrying that Mrs Pearce's feelings towards Amy and the serious difficulties in their relationship had apparently not been picked up while they were there. The staff remembered her as a capable, coping mother, and had not had any special concerns about her. As I continued working with Mrs Pearce, it seemed to me that a big part of the problem was that, because Amy had so nearly died, now that she was out of danger Mrs Pearce was furious with her.

This still left the question whether there was anything the ward staff might have looked for or done differently. We used this opportunity on the ward to discuss ways we could try to identify where difficulties between mothers and babies might arise. This discussion would not have happened had I not been working in two different parts of the organization, since it is unlikely the staff would ever have heard what happened after Amy left the ward. Indeed, this story is a clear illustration of how things can go wrong when there is no way to link what happens or what is known in different parts of a large organization.

CONCLUSION

I was originally invited on the ward because of the stress of the job. This is surely an understatement about what is inevitably involved in working with such raw and vulnerable babies. The staff must to some degree distance themselves from the babies and from their own pain in order to perform their tasks. The intense feelings evoked by being exposed to these infants must not be underestimated. An acceptance by managers and the staff that feelings can and need to be expressed is essential.

In thinking back on my work with this unit over a period of about eight years, I can see that I was used in a wide variety of ways to address these feelings. Sometimes I was involved with individuals, where my role was in helping them to separate out personal difficulties from professional ones, and to direct them, where appropriate, to find the right kind of help. At other times I worked with the staff as a group, exploring their relationships and how these helped or hindered the task. Always my function was to facilitate an awareness of the emotional issues on the ward: the inevitable grief, pain, helplessness and sometimes hopelessness. Greater awareness and understanding of these feelings, and allowing for their expression, led to better working practices and to a happier ward.

Chapter 7

Containing anxiety in work with damaged children

Chris Mawson

There are mental pains to be borne in working at any task, and these have to be dealt with by us as individuals, each with a personal history of having developed ways of managing or evading situations of anxiety, pain, fear and depression. Collectively, in our institutions, we have also learned to do this, installing defences against the painful realities of the work into our ways of arranging our tasks, rules and procedures. It is incumbent on us to try in whatever way we can to explore these aspects of our working practices, in order that our ways of coping do not grossly interfere, subvert or even pervert our efforts.

To understand the worlds of work occupied by ourselves and others, we need to be aware of the particular kinds of pain and difficulty encountered in everyday work situations. As Obholzer has observed: 'In looking at institutional processes it is obviously very helpful to have some inkling of what the underlying anxieties inherent in the work of the institution are. . . . Given a knowledge of the nature of the task and work of an institution it is possible to have, in advance, a helpful, fairly specific understanding of what the underlying anxieties are likely to be, even though one might not know the "institution specific" nature of the defences' (1987: 203).

WORKING WITH DAMAGED CHILDREN

Thus, when I was asked to consult to a child health team in a large teaching hospital, I anticipated certain difficulties. I knew they were involved in the assessment, long-term treatment and support of very young children who had been physically or mentally damaged from birth or soon after, and I expected from the outset to encounter considerable mental pain both in myself and in members of the team stemming from the workers' close contact with these damaged children. I anticipated, as was indeed the case, that the workers would frequently feel depressed, despairing of being able to make a worthwhile difference in the children's lives. I also expected that they would sometimes feel intensely persecuted by these feelings, even to the extent of experiencing at some level a measure of hostility towards the children themselves. It was likely that such intensely guilt-inducing feelings would often be deflected outwards and away from the work, in all probability finding their way into other parts of the institution, where they might well have adverse effects on working arrangements and interprofessional relationships.

In order to gain a real understanding of the team's experience of their work, I knew I would need to immerse myself in these experiences over a long period, as they shared them with me and with one another in our regular meetings. The following vignette, from one of my first meetings with the team, illustrates something of these problems:

Marie, a young physiotherapist in the Walsingham Child Health Team, described her visits to the home of a small child with a deformed hand. Each time she went, she knew her treatment would cause the child intense pain. It was clearly saddening for Marie to see the child freeze and turn away from her as soon as she set foot in the family home. She began to adopt a brusque and matter-of-fact manner with both child and mother, at times being quite aware that she was being cold and impermeable, but for the most part conscious only of a heavy sense of persecution and dread whenever she visited. She felt ashamed and defensive whenever she discussed the child and her treatment with other members of the team, and came to feel that this one case was casting a shadow over her enjoyment of her work. To protect herself against her guilt, she tried to tell herself that she was only adopting an appropriately professional distance, and that the occasional reproaches from the child's mother were really evidence of the mother's inappropriate need for closeness with Marie.

When this was explored in one of the first meetings the team had with me, there was a powerful reaction against opening up the issue of professional distance, and great resistance to the idea that it can be used to defend us against painful feelings in our work. It was as though the whole group felt attacked by me, and for much of the meeting I felt as if I were a sadistic person forcing an unwanted and painful treatment on them. They told me forcefully that they did not want me to make the pain of their work more acute, even if this was only a temporary effect.

It was clearly important for Marie to feel that her colleagues from other disciplines, particularly those whose role did not involve physical contact with the child, realized Marie's sense of hurt and rejection when faced with a child who was afraid of her, who did not perceive her as a healer or helper but as a cruel and sadistic figure who came into her home only to cause her pain. Initially it was very painful for Marie to talk about experiences which caused her so much shame and guilt. The wish in the team was to treat it as Marie's problem, which added to her stress and interfered with the whole team's learning from her experience.

Once the team became able to discuss these kinds of experiences in a setting where anxiety and guilt over feeling inadequate could be contained and understood, it was possible for us to see the sad irony that becoming defensively hard and impenetrable had in fact made it much more likely that the child would perceive the physiotherapist as sadistic. To work well with such children, and to be clear and supportive to their parents and families, professionals cannot afford to defend themselves by erecting these sorts of barriers.

PROVIDING A SAFE FORUM

Before such difficult feelings can be openly explored in a group, particularly when the members work together on a day-to-day basis, it is necessary to provide conditions of safety, respect and tolerance, so that anxiety and insecurity can be contained and examined productively. It is essential that a bounded space is created within which participants can begin to tolerate bringing more of their feelings than they are used to doing in other work activities, in an atmosphere which encourages openness and self-examination. Holding group meetings on the same day and at the same time each week helps strengthen this sense of containment, as does ending the meetings on time. It is not punctuality for its own sake that is important, but it is almost invariably disturbing for group members to feel that their emotions dictate the 'shape' and structure of the meeting, as well as its atmosphere and content.

The basic disposition of the consultant is important, too. The sense of security in the group is greatly encouraged by the consultant's restraint from judging and blaming, and 'knowing' too much too soon, or seeming to believe in quick solutions. It also helps if the membership of such a group is not constantly changing. The group often depend upon the consultant to stand up for the value of struggling for understanding, rather than rushing into the solving of concrete problems to get rid of uncomfortable feelings. They often find it useful to have such discussions in the presence of a consultant who is not a part of the organization, but this is not always the case.

Whether or not there is an external consultant, it is necessary for members to learn not just to listen to the content of what is brought to the discussion, but also to allow the emotional impact of the communications to work on and inside themselves. When primitive anxieties are stirred up, there is a natural tendency to try to rid ourselves of the uncomfortable and unwanted thoughts and feelings, locating them in others inside or outside the group, as described in Chapter 5. For example, recall how in the illustration given earlier I was temporarily experienced by the group as cruel, forcing on them an unwanted painful experience by looking at the issues in detail. When I was told that they were unsure they wanted such a painful 'treatment' if it made the pain of their work more acute, it was almost word for word what the parents had said to Marie. For a while, I had in turn felt in relation to the staff much as she must have felt with her young patient, saddened and guilty that my work was being experienced as cruelty. I had gone away from the meetings feeling somewhat persecuted, and had been tempted to defend myself by withdrawing from their reproaches and putting up something of a barrier, while telling myself that this was merely appropriate professional reserve. It was listening to my own feelings in this way that helped me to see how similar all this was to Marie's predicament. It was therefore possible not only to hear her feelings, but also to recognize from first-hand experience how such feelings are defended against, not only by her but also by the entire group in the institution. Understanding gained in this way can sometimes be put back to the group, or by the worker to the client, and, if timed sensitively, tends to carry a great sense of conviction.

In describing difficult work situations, members of the group will not only be communicating information, but will also be conveying states of mind which are often very disturbing and painful. From infancy we evolve the expectation that we can gain some relief from these pressures by seeking a 'container' for the painful feelings and the part of ourselves that experiences them. Partly, we unconsciously try to rid ourselves of them, but there is also the hope that the recipient of the projected distress might be able to bear what we cannot, and, by articulating thoughts that we have found unthinkable, contribute to developing in us a capacity to think and to hold on to anxiety ourselves. (These complex processes, termed projective identification, were discussed in detail in Chapter 5. See also Bion 1967; Klein 1959.)

SHOULDERING INADEQUACY

In many work situations, the chief anxiety which needs to be contained is the experience of inadequacy. The following example is drawn from my consultation to the staff of the Tom Sawyer Adolescent Unit, who were complaining about a difficult group of adolescents:

After several weeks of feeling increasingly useless as a consultant, inadequate and quite irrelevant to the needs of this hard-pressed group, I was told haughtily by one member that they would be better off without me. They would do better to organize a union meeting or an encounter group. I felt ridiculed, devalued and somewhat provoked. Another member of the team complained that I invariably took every opportunity to divert them from their real task. A third, speaking in falsely concerned tones and with knitted brows, asked why people like me were so intent on causing confusion by always looking more deeply into things. They were, after all, just honest workers whose only wish was to be left alone to get on with a difficult job, with little or no support. Yet another wondered why I bothered with them, and whether I was some kind of masochist.

Just when I had taken about as much as I could without losing my temper, another staff member, who up to that point had remained silent, said how despairing she had been feeling in her work lately, and how devalued. She felt her efforts had been under attack by some of the adolescent clients and their families. Another then added that it seemed their work was frequently undermined by the administrative staff who were supposed to be supporting them. It emerged that the whole team had been criticized recently by management for their handling of a difficult and sensitive situation in the unit.

It was at this point that I was able to make sense of my own feelings and the way I had been made to feel by the group. I could then put into words the team's deep sense that they and their work were under attack. In turn, they had needed to make me feel unwanted, ineffectual and under attack, partly to get rid of their own feelings, but also to show me what it felt like for them; this may have been the only way they were able to let me know. It extended to their trying to get me to give up on them, or else to retaliate. Just as they sometimes spoke of going home wondering whether they should resign, or whether or not to appear at work the next day, they had spent a month testing whether I would have the tenacity (or was it masochism?) to keep coming back to them. Another previously silent member confirmed this, saying she had secretly hoped that I would be able to keep going and not 'pack it in'. She also had wondered whether I had anyone to whom I could turn when the going got tough.

This led to a change of emotional climate in the meetings. It became possible to reflect on what had been taking place in the room and to make useful links to the current problems both in the team and in the wider institution. For example, it was possible to consider the predicament of some of the team's patients and families who, in extreme distress, often seemed to use the same projective mechanisms for alleviating their anxieties as the team had been doing with me. The feelings of the staff mirrored those of the parents, who had repeatedly been made to feel useless and impotent. When such feelings of inadequacy are unbearable the temptation to 'pack it in' can be too strong to resist, and this is precisely what had happened with many of the children there. Their presence on a psychiatric ward felt to them (and also to their parents) as evidence that the job of parenting them had become overwhelming and had been 'packed in'. The children had made the staff feel much as they had made their parents feel, and in turn the staff had made me bear the impact of these violent and demoralizing feelings. Furthermore, the question of whether I had my own sources of support could then be linked with the team's desperate need to find support and understanding in the face of such projections from their patients, so they would neither have to become masochistic nor have to 'pack it in' and resign.

The group came to feel that it had not so much been me who had been diverting them from their task, but that they had unconsciously been preventing me from doing my work with them. Their sense of having acted with some collective nastiness towards me made them feel guilty, but there was also the reality of what we had weathered and thus discovered together. This was of far greater value than any amount of abstract discussion or lectures – the latter having been suggested by them when free discussion had felt so bad and worthless. They had been able to experience someone who had obviously been buffeted by their attacks, but who had been able to contain feelings without hitting back or abandoning them. This demonstration of using reflection to manage feelings and reach understanding carried great conviction and helped them to move forward. At the next meeting it was possible for them to connect their fear that I would give up on them with their patients' anxieties that the staff would stop caring for them if they were too negative and unrewarding. They were also able to acknowledge their own fear that they would become too full of hurt and anger to continue their work, and that they really were at risk of abandoning their already traumatized clients. This had been mirrored in my impulses to explode or leave them, which I had managed to contain before acting on them.

Another common anxiety met by hospital workers is related to their inadequacy in the face of death; this is especially painful when it is a child or baby that has died. There is grief about the death itself, but also the feeling of having failed to save a life. The following example is taken from my consultation to the Walsingham Child Health Team:

> As we were arranging the chairs into a circle a booming voice could be heard just outside the door – which was still open because there were five more minutes before our starting-time – saying 'Is this a séance?' The voice belonged to Dr Royce, a senior consultant paediatrician who did not attend the meetings, despite having been invited many times. There was no apparent reaction, as though nobody had heard this comment. However, when the meeting began, it seemed to me unusually sluggish and half-hearted, with team members looking at one another for an instant and then breaking off eye contact. There were then a few remarks complaining about the lack of participation by medical colleagues, and why they didn't value the meetings.
>
> As I listened, I wondered what negative feelings about our work were being attributed to the 'absent profession'. I recalled similar remarks in the past about doctors' non-attendance: an often-shared attitude on the ward was that those who did not attend the meetings were commendably busy, while those who did had too much time on their hands. I also remembered that this had been a week in which the condition of several children on the ward had worsened, and a baby had died. There had been quite a subdued atmosphere before everyone had arranged their chairs, and nobody had made coffee today, which was unusual. I found myself thinking again about Dr Royce's jokey putdown. A séance is an attempt to contact the dead, and it suggests an unwillingness to face loss. Bearing all this in mind I decided to take up Dr Royce's remark, saying I had been surprised that not only had nobody commented on it, but there appeared to have been a concerted effort to act as though it had not been said. I wondered if they felt that their pain and loss could easily be denigrated.
>
> Alison, a physiotherapist who tended to permit herself closer emotional contact with the children than most of the others, then spoke of the difficulties in expressing feelings of grief in the hospital. Joan, an occupational therapist, spoke of her relief when a senior paediatrician had wept at the child's bedside. Alison remarked that nurses were labelled 'emotionally over-involved' if they

grieved, and others chimed in with complaints about the 'stiff upper lip' culture. There were a number of issues here, but what I chose to address was the way in which the group preferred at that moment to think of this repressive culture as belonging to the nurses, rather than as something in themselves. Only when the members could face their own 'stiff upper lips', and their conscious and unconscious equivalents of Dr Royce's mockery, would they be able to carry through the necessary work of mourning for the baby, and for the many experiences of failure and limitation represented by that loss.

This was a moving and productive discussion, but in spite of the obvious shared relief, I was left feeling doubtful about whether the lessons learned would be generalized and applied elsewhere. Perhaps it was only in that particular setting that professional defences could be lowered and such painful experiences explored.

CONCLUSION: CONDITIONS FOR GROWTH AND DEVELOPMENT

This raises a question about the potential for growth and development in groups, and how it can be supported. When painful work situations, such as those described here and in other chapters, are worked through again and again, it becomes possible for some degree of individual change to take place. Institutional practices can be scrutinized and sometimes changed, though this is rarely without difficulty and resistance. The 'change in emotional climate' mentioned above refers to shifts in the group from a highly defensive and mistrustful attitude towards one of regret verging on depression, as they recognized how efforts to protect themselves had led to treating others badly. The experiences I have described in this chapter stand out for me, not only because of the discomfort, but also because they are such vivid examples of the shift from a *paranoid-schizoid position* to one in which there was a preponderance of *depressive anxiety* (see Chapter 1). In the former position, the fear is of attack and annihilation, blame and punishment. Primitive defences against paranoid anxiety, if carried too far and with too much emotional violence, lead to the severance of contact with reality. For example, staff may deny the reality of the degree of damage, and of the limitations of what they can offer, as happened when the Walsingham team often felt under pressure to engender false hopes about the degree of improvement which could be expected in severely handicapped children.

The shift in emotional climate does not, however, result in freedom from anxiety. Instead, our fears of what others are doing to us are replaced by a fear of what we have done to others. This is the basis of genuine concern, but guilt and facing one's insufficiency are painful to bear. If these anxieties are not contained – and we therefore cannot bear them – there is likely to be a return to more primitive defences, to the detriment of our work and mental health, as was the case in the example of the staff grieving over the baby's death, where denial and mocking took the place of sadness and loss until the feelings could be worked through in the group discussions.

I have tried to demonstrate how important it is for staff involved in painful and stressful work to be given space to think about the anxieties stirred up by the work and the effects of these anxieties on them. The cost of not having this is considerable, both to clients and to workers. As well as offering much needed support, consultation can offer the opportunity for insight and change in the group and wider institution, *if* the pains and difficulties can be tolerated.

Chapter 8

Till death us do part
Caring and uncaring in work with the elderly

Vega Zagier Roberts

Caring for elderly people brings with it particular stresses, insofar as ageing is the fate of all who live long enough. It inevitably stirs up anxieties about our own future physical and mental decay, and loss of independence. It also stirs up memories and fears about our relationships with older generations, especially parents, but also grandparents, teachers and others, towards whom we have felt and shown a mixture of caring and uncaring. This chapter discusses how these anxieties were dealt with in one geriatric hospital. However, the processes described exist to some extent in all caring work.

THE INSTITUTION

Shady Glen was a specialized hospital for severely impaired elderly people who, without being particularly ill, required intensive, long-term nursing care. It had two wings: the smaller North Wing had three rehabilitation wards for those patients who were thought likely to be able to leave the hospital eventually; South Wing had four 'continuing-care' wards for those who were not expected ever to be able to live outside the hospital again.

The four wards of South Wing were particularly bleak and depressing. The beds were arranged in a circle around the edge of each ward, pointing towards the centre, from where the nurse in charge could keep a watchful eye on everyone. Squeezed between each bed and the next one stood a small wardrobe and chest of drawers; there was little space for personal possessions, and virtually no privacy. A few patients could move about with walkers, but the others spent most of their time in bed or sitting immobile in chairs. Most were totally dependent on the nursing staff for all their physical needs, and were fed, toileted and bathed on a fixed schedule.

The nurses maintained a high standard of physical care. There were few bedsores or accidents, little illness, and the patients were clean and well nourished. However, the managers of Shady Glen were concerned about the poor quality of life for the patients in South Wing, and asked the senior nurses of the South Wing wards to form a working party to explore what could be done to improve the situation. It quickly became apparent that patients' quality of life could be examined meaningfully only in conjunction with the quality of life for the staff working on the wards, and also that other significant hospital staff could not be left out of the project if real change were to take place. The working party was therefore expanded to include the heads of other departments providing patient treatment. Two external consultants were brought in to assist the working party in thinking about the stresses in the continuing-care wards, and considering how these might be coped with better. They were then to present their findings and recommendations in a report to the senior managers of Shady Glen.

STRESS AND INTERPROFESSIONAL CONFLICT

Morale among the nurses was very low, and relations between them and the other professional groups involved in the treatment of patients were antagonistic and competitive rather than collaborative. The nurses felt, not without some justification, that they were left to bear the brunt of the strenuous but thankless routine of physical care, unsupported and unappreciated. This kind of work has low status within the nursing profession – just as the patients on these wards could be said to have low status in society – and many of the older nurses at Shady Glen lacked the training and technical expertise needed for jobs elsewhere. They felt their wards were used as a dumping-ground for people that everyone else – doctors, families, society – had given up on and wanted kept out of the way, but well enough looked after that no one would have to feel too guilty about having rejected them. Not only did the nurses get little positive feedback from colleagues, patients or patients' families, but they got little inner satisfaction from the sense of a job well done. None of them felt these wards were a place where they would wish themselves or their loved ones to spend their last years.

The division of the hospital into two parts, one for patients who would improve, and another for those who would not, exacerbated the problem for both patients and staff. Many patients died soon after being transferred from North Wing to South Wing, as if they had received a death sentence. Staff on the continuing-care wards were deprived both of hope and of the satisfaction of seeing at least some of their patients improve and move back into the community. The alleged rationale for this division was that the two kinds of patients required different treatment approaches, and that the presence of 'incurables' would retard the progress of the less impaired patients, as if their condition were contagious, though there was little evidence for this.

At the same time, the nurses were not in the business of helping patients to die, as in a hospice, since most deteriorated only very slowly and remained on the wards for many years. It was as if the patients were 'on hold', the nurses just struggling against the gradual encroachment of decay. In the face of all this, any idealism or enthusiasm in newly arrived nurses was rapidly extinguished. New ideas they offered were rejected as impractical, or even sabotaged. As a result, those with ideas and choices rarely stayed long, and the staff from departments other than nursing tended to focus most of their efforts on the rehabilitation wards, adding to the continuing-care nurses' sense of being abandoned.

In the absence of the usual nursing goal of assisting patients to get well, the nurses did the best they could to keep patients as well as possible, which translated into keeping them safe: preventing accidents by keeping mobility to a minimum, discouraging the keeping of personal possessions which might get lost or stolen, keeping patients out of the kitchen in case they burned themselves. This policy, while depriving the patients of individuality and dignity, added to the quantity of work to be done by the nurses, so there were rigid schedules for meals, drinks, toileting and dressing in order to get it all done. Furthermore, since other professionals, like occupational therapists and physiotherapists, were mainly oriented towards increasing patients' mobility and independence, and since the services they offered tended to clash with ward routine, friction between the various disciplines was inevitable.

THE CONSULTATION

The antagonism between the nurses and staff from other departments was so great that the two consultants initially worked separately, one consulting to the senior nurses on South Wing and the other to the heads of the departments providing specialist inputs to the wards: speech, occupational therapy and physiotherapy. The plan was that the two groups would each first explore their own concerns and develop their own ideas for improving the quality of life on South Wing, and later come together to work on joint recommendations to make to management.

The nurses were at first apathetic and resistant to the whole project. They had worked on the continuing-care wards for a long time, were cynical about managers' implementing any of their suggestions, and were in any case sure that very little could be done, given the extent of the patients' disabilities. Everyone found their attitude very frustrating; even the senior nursing officer, who usually defended 'her' nurses from criticism from outsiders, chided them for under-mining the project.

In contrast, the members of the other group were young, enthusiastic and full of ideas. As heads of their own departments, they were accustomed to making decisions fairly autonomously, and for many weeks they worked eagerly at coming up with new programmes and plans for improving the quality of life on the wards. But the initial excitement gradually gave way to discouragement, as they anticipated – or actually encountered – the nurses' resistance to their ideas. Finally, the group became listless and work ground to a halt, everyone complaining, 'What's the point when they just won't co-operate?' The project had reached an impasse.

A chance occurrence some months into the consultancy changed this. Someone interrupted a meeting to ask for a patient's record, and it was revealed that many speech, occupational and physiotherapy records were months behind. This was the first time that any deficiency in the work of these departments was recognized. The group now began to work at reviewing their own services and improving them, rather than blaming everything on the nurses and focusing on how to make *them* change. They worked without the earlier excitement, but with more effect. At the same time, without there having been any formal contact between the two groups, the nurses became livelier in their meetings with their consultant, coming up with ideas of their own to contribute to the project. Within a few weeks, the two groups started joint work on what could now be experienced genuinely as a shared task, rather than a vehicle for apportioning blame. They drafted proposals for a new approach to continuing care, and these became the core of the consultants' report to management of their findings and recommendations (Millar and Zagier Roberts 1986).

THE REPORT

The central recommendation was to re-define the primary task (see Chapter 3) of the wards. Up to this point this seemed to have been to prolong physical life, keeping the patients in as good physical condition as possible for as long as possible. The proposal was that it should be 'to enable patients to live out the remainder of their lives in as full, dignified and satisfying a way as possible', which might or might not include their moving out of the hospital. This definition would mean that all the various professionals involved in patient care could see their particular work as contributing to a common purpose, rather than having conflicting and competing aims.

This change in task definition had major implications. It invited re-examination of practices previously taken for granted, such as the nurses' emphasis on safety as a priority, with its consequent depersonalization and loss of dignity for patients. Instead, the new aim required considering how to encourage such independence and autonomy as were possible, identifying differences between patients, so that some could make their own tea or leave the ward unescorted, even if others could not, and even if some moderate risk were involved (provided the patient wished to do so). This not only gave patients more self-respect and choices, but lightened the workload for staff and restored some meaning to their work. The greater dignity and sense of personal identity for patients if they wore their own clothes, no longer had wristband identification and had their personal possessions around them came to be regarded as outweighing the risks involved.

The new primary task definition also had implications for how the hospital was structured, that is, where boundaries needed to be redrawn. Boundaries delimit task-systems (see Chapter 3). Whereas before each discipline or department had had its own discrete task, and was therefore managed as a separate system, the new definition of a shared task required a new boundary around all those involved in patient care. (This is described in more detail in Chapter 20.) Furthermore, the separation of rehabilitation from continuing-care wards no longer had any rationale, since their previously different aims were now subsumed under a single task definition. By doing away with this, some hopefulness could be restored to the work.

Finally, the report recommended developing improved support systems for staff, particularly during the period of transition from the old way of working to the new. This is discussed further near the end of this chapter.

ANXIETIES AND DEFENCES IN INSTITUTIONS FOR INCURABLES[1]

The situation of severely disabled people who are neither dying nor likely ever to improve enough to leave an institution produces particular anxieties both in the residents and in those caring for them. Miller and Gwynne (1972) made a study of institutions caring for people with incurable, mostly deteriorating, physically disabling illnesses, but much of what they described is very similar to what was happening on South Wing. For the residents, entering this kind of institution is inevitably accompanied by a sense of having been rejected – by family, employers and society generally. Those inside such institutions are not necessarily more handicapped than those outside, but they have actually been rejected, if only by having no family to look after them, or no money to pay for care at home. Crossing the boundary into such institutions means joining the category of non-contributing non-participants in society: they lose any productive role they may have had, and with this, often, all opportunity to continue making decisions for themselves. Being treated differently from self-caring and able-bodied people, they experience great loss: 'I am no longer what (who) I was.' It is as if they are already socially dead, although they may be years away from physical death. The staff of such institutions can also have feelings of having been rejected and abandoned. Projective identification processes (see Chapter 5) can contribute to their over-protectiveness of the patients and their anger at patients' relatives and their own colleagues.

These were not the only difficult feelings which emerged during the consultation to Shady Glen. Others included staff members' anger at uncooperative patients and hatred of their failure to improve; discomfort with being still relatively young and healthy; anxieties about their relationships with the ageing members of their own families, and about their own ageing; and guilt for preferring some of their charges and treating them differently, while wishing they could be rid of some of the others, which could happen only through death.

Defences by the staff against becoming too aware of these disturbing feelings included depersonalizing relations with patients by treating them as objects, and by sticking to rigid routines; avoiding seeing common elements between themselves and the patients; illness, absenteeism and exhausting themselves to avoid feeling guilty. There was also an enormous anxiety throughout the care

1 Readers who have struggled to promote the personalization and dignity of clients and patients in institutions like those described here may object to the use of words like 'incurables' and 'inmates'. However, these stark terms, used by Miller and Gwynne in 1972, have been retained here not only for historical reasons but also to underline the harshness of the experiences being discussed, which can be glossed over by using more modern and politically correct language.

staff about being blamed. This probably arose largely from their internal and unconscious conflicts, but was attributed to their being held responsible for keeping patients safe and well. It produced a preoccupation with patients' safety, rigid routines designed to minimize the chance of making mistakes, and a hostile defensiveness towards colleagues and patients' relatives. The widely felt, but largely denied, doubts about the adequacy of the service contributed to the pervasive tendency towards blaming others.

TWO MODELS OF CARE

The anxieties inherent in any work give rise to institutional defences in the form of structures and practices which serve primarily to defend staff from anxiety, rather than to promote task performance. Miller and Gwynne (1972) identified two models of care in institutions for incurables, each involving a different central defence. The first, the *medical* or *humanitarian defence*, was based on the principle that prolongation of life is a good thing. This tends to be accompanied by denial of the inmates' unhappiness, lack of fulfilment and sense of futility. Inmates' ingratitude is an affront to these values. This defence produced what the researchers called the *warehousing model* of care, that is, encouraging dependence, and depersonalizing inmate–staff relations and care. A 'good' inmate is one who passively and gratefully accepts being looked after.

The second, the *anti-medical* or *liberal defence*, was based on the view that inmates were really normal, 'just like everyone else', and could have as full a life as before, if only they could develop all their potential. This defence produced what Miller and Gwynne called the *horticultural model* of care, defining the aim of the institution in terms of providing opportunities for the growth of abilities, while denying disabilities. There tends to be excessive praise for minor achievements, like the praise adults give for a small child's first drawings, and denial of inmates' failure to achieve social status. A 'good' inmate here is one who is happy and fulfilled, active and independent. Eventually, of course, nearly all of them fail.

It is easier to see the inadequacies of the warehousing model, but the other is also inadequate: the demand for independence may be distressing to some people whose physical and mental strength is declining. In many cases, they have been struggling for years against increasing infirmity, and some may give up this struggle with relief upon entering a nursing institution. Others want to continue to fight. These two types need different kinds of care and different attitudes in their carers. When models of treatment are based on defensive needs in the staff, however, these kinds of distinctions among different clients' needs may not be made, since they require thought and facing reality. Instead, one model is likely to be applied indiscriminately to all, on the basis of being the 'right' way to work, rather than as appropriate for the needs of a given individual at a particular time.

Both models represent unconscious psychological defences against unbearable anxieties aroused by the work, and by the very meaning of the inmates' having entered the institution. There is guilt about the social death-sentence that has been passed, and ambivalence about whether at least some of the patients might not be better off dead than alive. Similar splits occur in other institutions, for example, between cure and care in work with the mentally ill (see Chapter 13) or with the dying (see Chapter 10). In all these cases, care tends to be unjustly devalued, while cure is pursued against all odds.

MOVING TOWARDS INTEGRATION

Both the medical and the liberal defences were operating at Shady Glen, the first among the nurses, the second among the specialist therapists. Each group was unquestioningly committed to its own model. The therapists blamed the poor quality of life for patients at Shady Glen on the nurses' being unco-operative and too set in their ways to entertain new ideas. The nurses agreed they were resistant to the quality-of-life project, but insisted this was for good reason: no one else was placed as they were to realize the full extent of the patients' disabilities. They also felt hostile towards the more privileged staff who could leave work at 5 p.m. and did not have to dirty their hands with the 'real' work: easy for them to have these airy-fairy ideas! Only *they* behaved realistically and responsibly; it was thanks only to their disciplined care and unswerving routines that the patients had any quality of life, free of the bedsores, illnesses and injuries so prevalent in other geriatric care settings.

Each group had split off and disowned unacceptable parts of themselves, projecting these into the other group, who were identified with the projections (see Chapter 5). The therapists unconsciously counted on the nurses to attend to details, and so did not take responsibility for these, which led to the nurses' being all the more weighed down and having to stick to routine all the more rigidly. Similarly, liveliness and hopefulness were split off in the nurses and projected, defending them against guilt and disappointment, while the therapists became virtually manic in their planning. As a result of these intergroup projections, the nurses actually *were* rigid, and the therapists *were* inclined to be careless.

In the first phase of the consultation, members of the specialist group were excited and hyperactive in producing ideas and plans, the impracticality of which they blamed on the nurses – and the nurses accepted this blame. Over time, the euphoria associated with this kind of manic defence – that everything was possible, if only others would not stand in the way – gave way to angry helplessness and a listless feeling of being stuck. The recognition by the therapists of a shortcoming in themselves – small enough not to have to be immediately denied, but significant enough to provoke self-examination – led to their beginning to re-introject split-off parts of themselves, including responsibility for routine, and recognition of their own and also their patients' limitations. Taking back these projections not only increased their capacity for realistic work, but permitted them to value more the actual and potential contribution by nurses to patient care. Freed of the projections, the nurses were enabled to re-own hopeful parts of themselves, previously split off to defend against disappointment and depressive concerns, and to begin to relinquish some of their own obsessive preoccupation with routine. As each group became more able to value the other, less anxious about being blamed and therefore less prone to blaming the other, it became possible for them to think together about how to bring about improvements in the patients' quality of life, and thus also in their own.

THE NEED FOR SUPPORT

The recommendations to the management, to re-define the primary task and redraw the defensive boundaries between professional departments (see Chapter 20) and between rehabilitation and continuing-care wards, were designed to reduce the institutional splits which were impairing rather than supporting the quality of life at Shady Glen. However, since institutional defences arise in response to the anxieties inherent in the work, dismantling defensive structures requires providing alternative structures to contain these anxieties. The final part of the report, therefore, focused on ways of developing new kinds of support systems. These were of three kinds.

In the first instance, because of the stressfulness and strenuousness of their work, the staff needed their efforts to be recognized and valued, with explicit acknowledgement of work well done. They also needed more face-to-face contact with the hospital management to counteract their sense of being marginalized, rejected and of low status. This might be achieved through regular visits to the wards by senior managers, to review staff needs and the development of their new practices.

Second, staff needed a time and place where it would be possible – and actively encouraged – to reflect together on their work and how it was carried out. In small groups with continuity of membership, and with positive support from management, staff might then begin to acknowledge some of the unacceptable feelings aroused by their work: the fear, dislike and even hatred they sometimes felt towards their work and the patients, and the anxieties stirred up by the constant proximity to human decay. Otherwise, they could only defend themselves in the kinds of counterproductive ways we have seen. Often, just being able to face these feelings with colleagues can reduce the need for such defences. This, in turn, can lead to more effective task performance, which produces more work satisfaction, which further reduces anxiety: a benign cycle. (For further examples of this process, see Chapters 7 and 10.)

Finally, there needed to be mechanisms for inviting, considering and implementing ideas for change from everyone in the system, whatever their status (including patients and their relatives), so that everyone could participate in joint problem-solving and feel a sense of contributing towards a shared purpose. Such a forum could serve to support thoughtful institutional self-review and development, as described towards the end of the next chapter, in place of the rigid, stagnant working practices and entrenched intergroup conflicts which had previously characterized Shady Glen.

CONCLUSION

Dictionary definitions of care range from affection and solicitude, to caution, responsibility, oppression of the mind, anxiety and grief. Care staff can experience their work in all of these ways. Containing such a spectrum of emotions is psychologically stressful. Since ageing is the inevitable fate of all who live long enough, personal anxieties and primitive fantasies about death and decay add to the strains of looking after the elderly. The pressures to split positive from negative feelings are likely to be particularly acute. We have seen how at Shady Glen this splitting was exacerbated by the way the hospital was organized, with its divisions among the disciplines and between rehabilitation and continuing care.

In all caring work there are elements of uncaring. To be 'weighed down by responsibility' invites flight from the caring task, which can at times be hateful. Obsessional routines of care can serve to protect patients from carers' unconscious hate, from what staff fear they might do to those in their charges if not controlled by rigid discipline. At the same time, these routines can provide organizationally sanctioned ways of expressing hate of patients who exhaust, disgust or disappoint staff. Alternatively, all hate is projected, and the patients' hatefulness denied by seeing them as totally curable.

In his short but seminal paper, 'Hate in the Countertransference', Winnicott (1947) discussed the hate inevitably felt by psychoanalysts for their patients, and by mothers for their babies. He stressed that the capacity to tolerate hating 'without doing anything about it' depends on one's being thoroughly aware of one's hate. Otherwise, he warned, one is at risk of falling back on masochism. Alternatively, hate – or, in less dramatic terms, uncaring – will be split off and projected, with impoverishment of the capacity to offer good-enough care.

Winnicott's paper has given 'permission' to generations of psychotherapists to face previously unacceptable – and therefore denied and projected – negative feelings towards their patients. Indeed, to become conscious of such feelings has become a fundamental part of their training. Such permission – from within ourselves and from the environment – to acknowledge and own the uncaring elements in ourselves and our 'caring' institutions is crucial, both for individual well-being, and for the provision of effective services.

Chapter 9

Fragmentation and integration in a school for physically handicapped children

Anton Obholzer

There are several ways of assessing or attempting to assess the functioning of an institution and whether it is performing its assigned task satisfactorily or not. In the case of an industrial organization, parameters such as productivity or the amount of profit generated are generally regarded as good indicators of the state of the organization. In human service organizations, it is much harder to ascertain effectiveness and functioning, although certain criteria apply here, too, such as the degree of staff turnover and the amount of illness or absenteeism. As a consultant, however, I often have no access to such information, and I have therefore become preoccupied with finding other criteria by which to assess the state of the organization. Over time, I have found that some of the same criteria that are useful in assessing an individual's psychological functioning can be usefully employed in the assessment of institutions. Central to this is the degree of integration, internal and external, on which satisfaction from work and relationships depend. To illustrate this, I will describe my work with a school for physically handicapped children, as it shifted gradually from fragmentation towards integration.

THE FIRST PHASE

Goodman School served a wide catchment area, and places at the school were allocated on the basis of educational, social and medical assessment. The school had approximately eighty children on its books, ranging from under three to eighteen years old, most with severe physical handicaps, cerebral palsy being the most common. Apart from the head teacher, there were ten full-time and six part-time teachers, as well as two nursery assistants. There were also five physiotherapists, a full-time nursing sister and her auxiliary, a part-time speech therapist and five general attendants. The school also drew on the services of outside staff, including a social worker, a school psychologist and myself.

My original appointment to the school as a consultant child psychiatrist was for one afternoon (three and a half hours) fortnightly. The post had been in existence for some years, and I had had several predecessors. However, there was no job description, only a vague assumption that the school should have access to the services of a child psychiatrist. From the first, I thought that, given the shortage of my time and the large numbers of people involved, it would probably be most economical and effective for me to consult to the staff group with regard to their work with the children, rather than working directly with individual children. However, it soon became clear that this was not feasible. For a start, there were no formal staff meetings. Such meetings as did take place were called at irregular intervals, usually at lunch time, and were, according to hearsay, poorly attended, regarded as 'teachers' meetings', and used essentially to distribute administrative information.

In the entrance hall to the school there was and still is a prominent notice on display: 'All visitors to report to the office.' This was ostensibly to prevent unauthorized and curious visitors and parents from wandering about the school. It also, however, captured the spirit of the institution, and I for one was treated for a long time with suspicion as to my motive for being in the school at all.

On my first visit, I therefore found myself in the office being interviewed by Mrs Ryman, the head of the school, an eminently reasonable thing to do from her point of view, particularly as she had had no say in my appointment. She was very vague about my role, and there is little doubt that I was seen not as helpful, but as a potential threat to be warded off or placated. For a long time, my only contact was with the head and her secretary. Although at times Mrs Ryman would mention a child, it was mainly to tell me that there had been a problem, but it had now been resolved. Once her secretary telephoned the day before I was due to come and told me there really was nothing specific to discuss. It took a great deal of effort not to accept this invitation to enjoy a free afternoon instead of grappling with the problem. On another occasion, I was told, 'Please don't come today – we are having a problem,' making it crystal clear that my coming was seen as likely to compound the problem, rather than possibly being of help.

Towards the end of the first year, a member of staff would occasionally 'intrude', and we would discuss some problem they were having with a particular child. Very often we would then attempt to arrange a meeting at which all staff concerned might meet in order to discuss the child in question. This often foundered on a difficulty of 'timing', particularly when it came to bringing staff from different professions together. The time-table was arranged in such a way that it was almost impossible for teaching and non-teaching staff to meet. Not surprisingly, relations between these two groups were very poor. What were consciously presented as time-tabling problems actually represented interdisciplinary conflicts and rivalry; the time-table was an unconscious institutional defence to prevent these groups ever meeting to work together. None the less, it became increasingly apparent that there were a number of staff who would welcome meetings to discuss issues of common interest, and eventually one was arranged.

DEVELOPMENT OF THE CONSULTATION

There is little doubt that the eventual institution of regular meetings for all staff was caused by the head's decision to retire early: in the ensuing vacuum, new opportunities for change were created. In the months leading up to her retirement, meetings with staff of several disciplines began to take place in the office more frequently. The notion that the children presented were not only there in their own right but as 'representatives of an institutional process' became more acceptable and understandable to the staff group. For example, a child presented for having stolen while the class went swimming was seen not only as an individual problem, but as a pointer to the issue of stealing in the school.

It was decided that regular staff meetings, to which I was invited, would be instituted at fortnightly intervals. It had taken three years for regular staff meetings to be instituted and for me to be invited to attend. Even then, the initial contract was for one term only, but this was subsequently re-negotiated to continue long term. Although I recognized the anxiety and expectation in the staff group concerning the coming meetings, I perhaps did not sufficiently acknowledge my own, but instead dealt with it by arranging with the school for a colleague to come with me.

My colleague and I wanted to adopt a model of being essentially non-participative but engaged in listening and attempting to understand the specific issues that were consciously and unconsciously preoccupying the staff. Our assumption was that an understanding of these processes would further their work. It was thus essentially a psychoanalytic model based on containing anxiety and searching for understanding which could then be used to inform decision-making. Not surprisingly, this model, though unspoken, proved too threatening and had to be modified. What was 'expected' of us was to participate specifically in our roles as child psychiatrists to the school. As such, we engaged in discussion as colleagues and fellow-workers, albeit in a different role and with different competence and experience. In addition, however, we would sometimes stand back and make 'consultative' observations on the interaction between ourselves and others. For example, 'Look at us, the way we're carrying on. I'm talking as a psychiatrist, you as a teacher, and neither of us seems to be able to listen to or understand each other.' This might then be further linked with professional preconceptions and the difficulty of interprofessional co-operation.

It is, however, not entirely accurate to say we were expected to behave in our original roles, because even that presented a threat. The least threatening position would have been for us to be 'one of them', and to behave like them. At the very first meeting, we were immediately asked what our topic was and what we were going to teach on that day – a model of teaching or being taught being what they were most at home with. It took some time for us to be accepted in our roles as being 'different' from them.

Over time, we attempted to capitalize on the fact that there were two of us. We did this by having one of us engage in the discussions from a professional stance, as child psychiatrist, while the other would adopt the role of institutional consultant and comment on the processes active in the staff meeting at the time. It took about a year from the establishment of the staff meetings for us to work as consultants to the institutional process as envisaged in our original model, and when my colleague left to have a baby, I was able to continue the work on my own.

THE ANXIETIES INHERENT IN THE TASK

Although some children at Goodman School were lightly handicapped, the majority had a substantial physical disability. Many were in wheelchairs, and orthopaedic devices were common. Some children could not even feed themselves, and were totally dependent on the staff for all caring functions. In some cases, the children suffered from progressive diseases, their condition deteriorating until they died.

Teachers generally understand their job to be helping pupils to acquire and use skills, developing their competence, and guiding them towards applying their learning so as to take up their roles in society later as independent adults. In this school, the teachers had to face the fact that much of what they taught would not be put to use, either on account of the deterioration, and sometimes death, of the child, or on account of the difficulty in integrating a disabled person into our present society. Similarly, the physiotherapists had to work with clients in whom there was seldom the sort of improvement or success that could be achieved in more acute conditions. In many cases, it was a matter of presiding over and slowing down as much as possible a process of deterioration. Even when there was a good and hopeful working relationship, the physiotherapists often had to witness the child leaving the school and their work being lost on account of the difficulties in continuing treatment outside, such as shortage of transport and treatment facilities, or lack of follow-up.

When there is so much pain inherent in the work, it is clear that some defences are necessary for the staff to remain in role and carry on with some hopefulness. However, at times the extent and nature of the defensive processes is such that they not only interfere substantially with the primary task of the institution, but they are also ineffective in their intended purpose of protecting the staff from the pain of the job.

THE INSTITUTIONAL DEFENCE SYSTEM

Defence mechanisms can be divided into two categories – personal and institutional. These are interrelated, partly because staff members with individual defences 'fitting' those of the institution are more likely to remain, while those whose individual defences are out of tune with the institution leave. It is worth commenting on the fact that several members of staff spoke about either having or having had a deformity of a physical nature, or of having siblings or children who had physical disabilities. It is not clear whether this was because there was a higher incidence of physical handicap in the histories of this group of teachers as compared with other teachers – and that teachers with this sort of history were more likely to be drawn into this work – or whether ongoing contact with physically handicapped children made them more aware of their family histories. There is probably a link between choosing to work in this type of institution and personal history.

There is a further link between personal and institutional defences. Each individual member of the institution – and his or her defences – comes to represent aspects of the whole. The teacher whose personal defence is 'I treat them as normal' not only represents him- or herself, but also becomes the spokesperson for that part of everyone else in the institution that needs to regard the children as normal, and of that part of the institution itself that deals with abnormality by denial. When there is a 'good fit' between the personal and the institutional defence systems, neither will be challenged: reality-testing does not take place, and the defensive process continues.

The task of consultation at Goodman School was not only to name the defensive processes, but also to attempt to identify the underlying basic anxieties and their present manifestations. A model of school functioning based on a denial of the children's handicaps made it difficult to help the children towards an acceptance of these handicaps, and for this acceptance to become the foundation on which their transition from school to work and adulthood might be built. For example, Rodney, a twelve-year-old boy with severe epilepsy, planned to be a bus driver. The acceptance of his epilepsy would have required his making a change of career choice; continued denial of his problem would make his transition from school to work a difficult one. School-leaving regularly created enormous problems for the leavers, but the pain of facing children's limits was avoided by the staff.

The need to deny the implications of the physical handicap was also supported by the way children entered the school. The entry procedure had traditionally been administered by outside agencies, separating the medical, social and educational aspects of the child's situation. Only at a late stage did the parents and child have contact with the school; they then usually met only with the head and focused essentially on academic matters. This process of induction, based on denial of the physical handicap in word and deed, served the defensive needs of both staff and parents. Later, when the child's growing up or deterioration inexorably brought the impact of the physical handicap to the fore, the relationship between staff and parents often deteriorated into mutual blame because the social and developmental implications of having a serious disability could no longer be ignored.

Meanwhile, parents in general were becoming more vociferous in their demands, more questioning of teachers and schools. As well, there was a growing movement away from hiving off children with special needs into specialized institutions, and towards integrating them into ordinary schools. These factors contributed to a situation where, if Goodman School were to survive, it had to provide the service it set out to – and it had to retain the co-operation of the parents. In the past, this had always been based on collusion between staff and parents, whose defensive needs were similar. Now there arose an increasing number of situations in which there would be protests from parents about the long-term costs of the traditional approach to working with these children, often followed by transfer of the child to another school.

Thus, the defensive processes not only did not serve to protect the staff very effectively from the pain of the work, but now also threatened them with the loss of jobs and the closure of the institution. As for the children, the damaging effect of abrupt changes of their environment can be readily imagined. It therefore gradually became clear to everyone that the defensive processes were anti-task, and that they urgently required review if the school were to survive at all. The role of the consultants became increasingly accepted, and staff meetings became a weekly event. Initially they alternated between 'administrative' and 'other', with the consultants continuing to attend on a fortnightly basis, but this fostered a split between action and understanding. To overcome this, the consultants then attended weekly, making links between issues arising in the two sorts of meeting.

SPLITTING AND FRAGMENTATION

The weekly staff meetings were open to all members of staff: in practice they were attended by the head, deputy head, most of the teachers, most of the physiotherapists, the speech therapist and ourselves. The cleaning staff did not attend, being engaged in cleaning at that time; nor did the school helpers, who were taking the children home during the time of the meetings. The size of the group fluctuated between eight and twenty.

Initially, the mere fact of the meetings being convened led to greater communication between individual members of staff and between different professional groups. Paradoxically, it became clear why staff meetings had hardly ever been held before – except to distribute administrative information – for they brought together various split-off aspects of a problem and initially resulted in increased distress. As put at an early meeting, 'It is good we are re-thinking issues and problems with children, discussing our problems with members of other disciplines, but when we are all in the same boat, we all feel helpless about the same child.' Staff meetings were clearly in the interests of the children, but were painful for the staff. The tendency was therefore for defensive splitting to take precedence over integration and its attendant pain. Splitting and denial (see Chapter 1) are among the most commonly used defence mechanisms in institutions. At Goodman School, they occurred between the school and the parents, the school and outside agencies, and within the school itself.

Splitting between school and parents

Splitting between school and parents was a common occurrence, and parental disturbance was often cited as a major source of difficulty, whether in communicating with a parent, or when a child was in trouble. If the parents were to blame, then by implication the teachers were not, and could therefore do nothing. An aura of helplessness would set in, everyone feeling stuck, and the child was not helped.

A good example of this sort of splitting was the case of Mrs Langham, the mother of a child at the school, who wrote a newspaper article about the increasing educational and social deterioration of her daughter, suggesting that the school system was to blame. Although the article was fairly reasoned, there was little doubt that the mother herself was in desperate pain about her child. At the staff meeting following the publication of the article, there was uproar. There was a point-by-point dissection of the article and several inaccuracies were pointed out. Then there was a discussion of the child's family history, with particular emphasis on the parents' neurotic disturbances. Nowhere was there any acknowledgement of the fact that the article raised issues vital to the school, nor any awareness of the possible underlying dynamics of the situation. The entire initial exchange thus took place within the paranoid-schizoid position (see Chapter 1).

With time, it became possible for us to interpret the denial that Mrs Langham's concerns and doubts were also held by many of the staff. It then became possible for the school to relate more sympathetically with the Langham family, but also for them to look at the issues raised, and to think constructively about families in the grip of similar dilemmas. At a subsequent meeting, a teacher spoke of another parent as a 'neutron bomb just about to go off'; the staff were able to consider ways of intervening before the situation got out of hand.

Splitting between staff and outside professionals

There were frequent complaints that the children often went to clinics or specialists of one sort or another, but that no reports were ever sent to the school. There were endless discussions concerning the need for better communication with outside agencies, and plans to write letters to certain regularly used clinics and specialists requesting information, as if the staff believed 'If only they told us what was going on there would be no difficulties.' They held tenaciously to this belief, despite their experience that when they *did* get reports, they generally found them of little value, adding minimally to what they already knew. They also expressed indignation when outside agencies involved with the children did not seek their opinion.

Here again the issue was one of splitting, in which the 'helpful information' was either held outside and not passed on, or else held by the staff but not requested. The depressive position (see Chapter 1) counterpart of the above view would include an acknowledgement of the difficulties involved in working with the handicapped, and an acceptance of the fact that nobody knew the answers or had a simple solution. Recognizing this led to greater sharing of difficulties and uncertainties, moving towards teamwork rather than competition and blame.

Splitting personal and professional parts of the self

Splitting up competencies among the disciplines was a common occurrence. For example, at one staff meeting there was a discussion about showing the children a film on epilepsy, which parents might also attend. When I asked who would lead the discussion, the answer, given as if to a stupid question, was 'The school doctor, of course.' Undoubtedly, the school doctor had a major contribution to make. However, the 'of course' answer also highlighted a splitting process in which staff insisted on regarding epilepsy as a medical problem. The result was that the many years of experience of other members of staff in how to deal with epilepsy were lost. The split safeguarded the non-medical staff from discussing a painful area in the lives of many of the children and their families. Yet it was precisely this non-medical discussion that might best help the children and their parents with the process of accepting the disability and integrating it into everyday life. Here again, the splitting process protected some members of staff, but interfered with the overall work of the institution.

A similar split arose regarding helping children at puberty to develop their sexual identity. Our observation was that the physiotherapists had a central role in this. What could professionally be perceived as hydrotherapy to enable a child to have freer limb movement was, at another level, an opportunity for close bodily contact and discussion of the feelings aroused by puberty. However, the teachers stated categorically that the physiotherapists were not equipped for such discussions, and suggested that a lecturer be invited to bring a film on sexual development. Although this might well have been of interest, it was not a substitute for the discussion that might have taken place in the ongoing physiotherapeutic relationship, one that had often developed over several years. This splitting of professional and personal parts of the self led to the loss of a wealth of personal experience acquired by the physiotherapists (and other staff) in reaching their own bodily and sexual adjustment. Although personal privacy needed to be safeguarded, there was little doubt that if the professionals could draw on this personal experience and not split it off from their professional expertise, they could greatly facilitate the adolescents' coming to terms with the changes their bodies were undergoing.

Similarly, when the staff were able to stay in touch with their personal experience of being parents or siblings, they had no difficulty at all in sharing the dilemmas of the children's families and hence in being more understanding of their reactions, rather than complaining, 'How can they behave like that?' When staff spoke of their experiences in taking children on outings or on holiday, they often marvelled at the patience and fortitude of the parents, yet this experience would soon be 'forgotten' when they returned to school, and the old complaints would resume.

INTEGRATION AND DEVELOPMENT

As a result of growing integration within the staff group, and diminishing the split between the personal and professional parts within individuals, new ways of working began to develop. One of the first was a physiotherapy group for new children, where they were allowed and encouraged to examine their own physical strengths and handicaps, thus developing their own body image. However, the group had been going for some time before the physiotherapists told the rest of the staff about it. As one of them said, 'At training school, we were always taught never to mention the disabilities, and to instead develop exercises that would require the use of the handicapped limb.' This supported the institutional defence system, based on denial, while reducing the physiotherapist role to body manipulation. It took time and support to enable the physiotherapists to break free of this and develop a new and more effective approach, one incorporating mind and body for both child and worker.

Similar developments took place among teachers, so that, for example, one teacher who two years previously had spoken of his role as 'only a teacher and that's it', transformed some classroom sessions with his children into wide-ranging general discussions. This was not an avoidance of the task of teaching but rather an augmentation of it, in which a personal component was made available in furtherance of the school's work. Yet, here again it took some time for the teacher to make his new approach public. As with the physiotherapists, the change conflicted with his teacher training, and he therefore worried it might be considered unprofessional.

At the organizational level, too, support for questioning old assumptions led to change, for example, regarding admission procedures. The school clearly had the potential and the knowledge to develop its own system of intake, better geared to the real needs of children, parents and staff, but this process was constantly slowed down by their reluctance to question that the old practices were the 'right' way, and any deviation from these were 'wrong'. Only gradually, through the staff discovering again and again the defensive nature of many of their traditional ways of working, could these be questioned, reviewed and, when necessary, changed.

Although the shift from the paranoid-schizoid position to the depressive one has been described as if it were steadily progressive, it is by its very nature unstable. Whenever anxieties increase, defensive processes are fuelled and there is a tendency to return to splitting. This oscillation, with pressure to shift away from the depressive position, means there is repeated loss of the capacity to face painful reality, guilt and concern. There is thus a need for the ongoing containment of institutional anxieties to safeguard depressive functioning. (For further illustration of the need for and usefulness of containment, see Chapter 7.)

CONCLUSION

In this chapter, I have attempted to draw a parallel between the paranoid-schizoid/depressive position spectrum in individuals and in organizations. The chapter illustrates how the functioning state of an organization can be assessed by monitoring measures such as splitting and projective identification, and how – just as for individuals – a containing intervention can shift the functioning towards the depressive end of the spectrum.

None of the institutional processes mentioned in this chapter is unique to this school. They occur in other schools, in other 'people' institutions and in society at large. For staff to function to the best of their ability, they must have an external and internal framework that allows for a sense of security – a security that can be used as a base from which to explore personal and institutional issues. A regular time-table with clearly defined staff meetings is essential to bring together all the groups involved in providing care. It is my belief that an outside consultant to the staff meeting is invaluable in making sure that such meetings do not become infiltrated and bogged down by institutional defence processes. This consultation must be managed so that it is supportive of the institutional management structures, neither in collusion nor in collision with them.

It is only with the provision of a containing environment that the institution can settle down to working at its task. Members need time to get to know each other and their roles in a task-oriented setting; 'chats' during coffee break or lunch-time are not sufficient, as they invariably shirk the most difficult issues of the day. It is only with time and ongoing work that staff can reach the important stage – personally, professionally and institutionally – of having the freedom to think their own thoughts, as opposed to following the institutional defensive 'party line'. Only then will they be able to develop their own style of work, and contribute fully to the task in hand.

Chapter 10

Working with dying people
On being good enough

Peter Speck

For most people, death only enters their lives a few times. When it does, it can precipitate a significant crisis as the individual seeks to adjust to the impact of the event. But what of the professional who in some settings faces death almost every week, if not every day? How does he or she cope with this abnormal exposure and still manage to maintain a professional role? Death is universal and will come to us all and to those we care about at some time. We know this intellectually, but we may well try to defend ourselves from the emotional impact of personal death or the death of someone close to us.

STRESSES OF WORKING WITH DYING PEOPLE

Working intimately with people who are dying can put one in touch with personal loss; unresolved feelings and anxieties may be evoked by the death of someone we are caring for professionally. Attempting to suppress or deny the personal impact can be stressful, leading to fatigue, sickness, compensatory over-activity, loss of effectiveness at work and at home, together with other symptoms often referred to as 'burn-out'. Attitudes expressed by others may add to this excessive stress. People may make comments such as, 'Oh, those poor people! All that suffering! You must have to be quite hard, or very dedicated.' Or, 'I don't know how you do it, day after day. Isn't it all terribly depressing? I mean . . . they die.' Comments such as these make explicit an expectation people have that the carer must be an extraordinary person offering perfect care. If the carer knows that the reality can often be somewhat different, then the expectation can be a source of stress. If the carer shares the expectation, then this gives rise to stress, too. One often shared expectation is that good care will make for a 'good' death. But death is not just sad or beautiful; it can be ugly, painful and frightening.

One of the unconscious attractions to working with dying people is that the work-role can serve to maintain the fantasy that death happens only to other people. When this defence breaks down because the work situation is too close to one's own, there is a real risk that one may be disabled from working – or staying in role – at all, as the following vignette illustrates:

In my role as chaplain, I was called one morning to the hospital intensive care unit (ICU) to see the parents of a dying boy. John, aged thirteen, had been knocked down by a lorry on his way to school. His mother, Mrs Brown, explained how he had been late for school and she had shouted up the stairs, 'You're late! Get out of bed or else . . . !' John eventually ran downstairs, dressing as he made for the front door. He said that he was too late for breakfast, ran off down the road, and at the corner ran out into the path of the lorry. He had been diagnosed as probably brainstem-dead, and the final definitive tests were awaited.

As Mrs Brown talked, I became more and more upset and unable to listen to her. I just wanted to shut her up and to get away. It was increasingly difficult for me to stay in the room, let alone in my role. I realized I needed to leave and sort myself out; eventually I apologized and left the waiting room. Outside, a senior nurse, René, asked me what was wrong. I said I was very distressed about David. 'David who?' she asked. I said, the David who was in ICU and brainstem-dead. 'But we haven't got a David,' she said, 'Only a John. Come and see.' She took me into the side room where John lay. I realized it was not David, my son.

Then I remembered how that morning, as I was leaving for work, I had heard my wife shouting up to our son 'You're late! Get out of bed, now!' When Mrs Brown had used almost exactly these words in telling me her story, she momentarily became my own wife, telling me that our son was dying. In the ICU waiting room, I had been so identified with the situation that I became a stricken father, not a chaplain. Once I realized what was happening, I returned to the waiting-room. As I apologized again for leaving, Mrs Brown threw her arms around me and said, 'It's all right, Vicar. I could see you were upset.' We were then able to work together on how she and her husband might face the impending death of their son.

Through over-identification, my sense of reality and my professional identity had both broken down for a while. After an experience such as this, it is tempting to try to defend against its ever happening again, and to make sure we are no longer touched by such events. But in the field of terminal care, identification will always be a possibility: there will always be patients who resemble us or someone who is significant to us in ways which stir up anxiety. We are repeatedly put in touch with past losses and reminded of the certainty of future ones. Each time, we are confronted with the reality that our work does not confer any special protection against death.

DEFENCES AGAINST THE FEAR OF DEATH

Identification, the consequent loss of boundaries and the difficulty of managing oneself in role in the face of death can lead the professional to develop strategies to minimize the possibility of this happening – conscious and unconscious defences against the emotional impact of the work.

Avoidance

Some people, for example, may avoid the difficulty of talking about death in a personal way by rationalization (the patient doesn't really want to know . . .); or by intellectualization (talking in terms of statistics); or by hit and run tactics (telling the patient bluntly and then leaving, avoiding all further contact). Others may try to avoid direct contact with death altogether.

A senior ward sister on a busy male surgical ward became quite hysterical one morning after a patient died while she was standing at the bedside. It was only the second time she had been present at a patient's death since qualifying over twenty years earlier. As a student nurse, she had witnessed a very traumatic death. When she had become distressed, the then ward sister told her, 'Pull yourself together and go and lay him out, then get the rest of the beds made.' She had vowed that she would never put herself in that situation again. Once she qualified, she always ensured that any dying patient had another nurse assigned so that she could avoid getting involved. She had managed that quite successfully until now.

Task-centredness and aggressive treatment approaches

The tendency to defend oneself by adopting a task-centred approach is not new. The priest can hide behind the ritual of prayers and sacraments, avoiding interpersonal contact; the doctor can use a stethoscope to silence the questioning of a patient. Such defences are often a way of protecting oneself from death, and of reassuring oneself that 'It won't happen to me.'

For clinicians, another way is to move into aggressive treatment approaches with techniques which may prove beneficial in some cases, such as radical reconstructive surgery, heavy doses of radiotherapy or highly potent cocktails of chemotherapy. The focus in such approaches is on potential cure, the 'breakthrough', which can confer a great feeling of omnipotence on the clinician. Such a disease-focused approach may be accompanied by a distancing from the patient as a person, and any patient who dies represents failure. Meanwhile, the human caring aspect is vested in the nurse or other paramedic who remains close to the patient and is identified with all that is good, loving and nice. When the patient dies, the nurse may become the recipient of the doctor's negative feelings because of the unaddressed rivalry that has developed between them. The rivalry may relate to a covert competitiveness for the gratitude of the patient, who may seem to have shrugged off dependency and asserted his or her autonomy in the only way left – by dying.

An attempt to address this split is seen in the developing speciality of *palliative care.* Here a partnership is sought between the doctor, patient and nurse, with both doctor and nurse involved in a personal way with the patient for whom they are caring. Surgery, radiotherapy and chemotherapy may still be employed, but the focus is shifted from cure to symptom control and pain relief.

CHRONIC NICENESS

Staff who work in hospices, which specialize in care for the terminally ill, usually appear to be cheerful people who work extremely hard to provide the highest standard of care for their patients and patients' families so that the dying can achieve a 'good' or 'nice' death. While there is little doubt that hospice staff are caring and dedicated people, one of the dangers which face them, and others who work long term with dying people, is that of 'chronic niceness', whereby the individual and the organization collude to split off and deny the negative aspects of caring daily for the dying. There is a collective fantasy that the staff are nice people, who are caring for nice dying people, who are going to have a nice death in a nice place. This protects everyone from facing the fact that the relationship between the carers and the dying can often arouse very primitive and powerful feelings which are disturbingly not-nice.

Having relinquished some of the traditional defences described above, carers can fall victim to this chronic niceness, as if acknowledging any negative feelings and thoughts about patients or colleagues might threaten the unity they have established. In order that everyone can continue to be nice to each other, the not-so-nice feelings get split off and displaced outside the staff group. For example, there may be much complaining about managers who 'don't understand the pressures . . . are always demanding more and more of us . . . don't seem to value what we do'. This can, of course, be true, but it can also indicate that the group has moved towards a paranoid-schizoid position in order to retain staff cohesiveness (see Chapter 1). In this case, whatever management may intend will probably be misperceived by the work group. Similarly, the split-off negative feelings may be projected on to the patients' relatives, who are then perceived as hypercritical of the standard of care and of the way in which the staff are looking after the patient. The staff's image of themselves as competent, caring and nice people may seem to be under attack, and it may feel like it is only the patient – who is already in a dependent relationship – who really appreciates what they are doing.

Survivor guilt and the need for gratitude

In my encounter with Mrs Brown described earlier in this chapter, I experienced a large measure of relief in finding it was not *my* son who was dying in the ICU. This feeling, which psychoanalysts refer to as 'manic triumph', often arises in the course of caring for dying people and brings with it a sense of guilt (Freud 1917). It can be difficult to acknowledge to oneself, let alone to others, that one feels pleasure in knowing that it is not one's own death or the death of someone important to oneself that is about to happen. In the face of the obvious distress of those who *are* experiencing death, one may then also experience guilt for having survived. There may be a strong desire to split off and deny such painful feelings. However, if one can contain simultaneously contradictory feelings of concern and relief, it is possible to tolerate the pain without resorting to excessive splitting, projection and denial (Klein 1959; see also Chapter 1).

Patients' gratitude can do much to alleviate survivor guilt. But patients are not always nice and appreciative, and carers will then have to find ways to deal with the negative feelings this arouses if they are to continue to manage themselves in role, as the following vignette illustrates:

Robert was a 35-year-old married man with two very young children. The day he received his diagnosis of cancer of the pancreas he was initially very distressed, and then withdrew and became uncommunicative. In the course of my routine visit to the ward, the staff told me they were worried about Robert and I agreed to visit him. At this first meeting, Robert made it clear he was not religious, and that the events of the past few days had done nothing to change this. I explained that I was not in the business of 'arm-twisting' religion, but was available to listen to his feelings about the events of the past few days.

Robert remained silent for a while, and then became tearful. Eventually, he said, 'It's not being able to see the kids grow up . . . but I can't talk about it now.' I said I recognized that he might not feel ready to talk at present, but asked for permission to visit again, explaining that staff could contact me at any time. Over the ensuing weeks of treatment and care, I visited Robert regularly. There was no formal religious ministry during these meetings, but we talked about what it meant to him to have an inoperable tumour, and about the impact of the treatment on Robert and his family. Some of this he was able to share with other staff as well. Robert expressed his grief over the loss of a future, his concern for his wife and young children, and his anger at what was happening. Once he discovered that I was concerned to relate to him *where he was* in his understanding of life, Robert was able to use me as a support and resource.

Then the time approached when I would be away on annual leave. I tried to prepare Robert for my absence, and introduced a colleague who would be available in my place over the holiday period. At this point, Robert developed a blocked duct, and had to have some palliative reconstructive surgery. Although he made a good recovery from the operation, it was clear his condition was deteriorating. His anxieties concerning his children became more acute. Although there had been several family sessions when the children had all been able to talk about the impact of their father's disease and probable death, there was much that Robert still wanted but felt unable to say to them. Following a suggestion from me, he decided he would write letters to each of his children and to his wife. The letters took a long time to write, but eventually he finished them and gave them to me for safe-keeping, telling his wife what he had done. He then asked me to promise to visit him the night before going on holiday.

When I made that visit, Robert was reasonably comfortable and wanting to talk. He reviewed much of the ground we had covered over the previous weeks, and also highlighted the areas which he still needed to explore. I felt Robert was trying to face the ending, but at the same time indicating there was much to be done and not much time. He expressed gratitude to me for the work we had achieved, but the overriding feeling was regret for all that had not been done. At this point I began to experience feelings of guilt, of having let Robert down by not being available for this new work, and a sense of failure for not having helped Robert achieve what he wanted in the time available. While I was trying to make sense of these mixed feelings, Robert suddenly asked me if I were going on holiday tomorrow. Then he asked if I were taking my wife and children. Again I said yes. There was a silence, and then, loudly and angrily, Robert said, 'Well, it's bloody well all right for you isn't it? I'll be dead when you get back. Have a bloody good holiday!' He then turned over in bed, away from me, and pulled the sheet over his head.

I felt very angry at this 'unfair' outburst, and for being made to feel guilty for going on leave. I felt unsure whether to tackle the issue with Robert or to leave and risk that Robert might later feel bad about his final farewell. At that moment, too, a part of me wanted him to feel bad for 'spoiling' our good relationship in this way. It was proving difficult for me to retain any capacity to think through what was happening. After a moment I said, 'I'm sorry we are parting in this way, Robert. I appreciate you're very angry with me for going, but that's the way it is. I must now say goodbye,' and I touched his arm as I said this. There was no response, so I left the bedside and went to find the ward sister. I told her I needed to talk through what had happened because I felt very angry. We went to the office, had a cup of coffee, and I tried to understand what had happened between Robert and me.

The need for support and containment

It is understandable that dying people should expect their carers not to desert them. The chaplain, especially, is expected to be available and nice at all times, remaining calm, pleasant, able to receive and contain anything 'dumped' on to him. This expectation is often shared by the other care staff of the institution which employs the chaplain. When staff feel stressed and fragile, it can be very useful to have a chaplain on board to be the recipient of negative projections. In this case, having managed my difficult feelings as best I could, I needed support myself, which I got from the ward sister.

As we talked things over, I realized that some of my anger was at Robert's illness and the inability of any of us to 'put it right' and restore him to health. I also recognized that I felt guilty not only for leaving, but also for not having worked more on Robert's anger earlier. I had thought we had addressed it, but now I experienced a strong desire to 'turn the clock back' and do it better, which is a characteristic aspect of endings. My feelings mirrored Robert's: a mixture of gratitude for what had been good, sadness and anger, and also guilt and anxiety about not having done all that might have been possible. By being able to ventilate some of my feelings with the ward sister, I came to see that for some time I had not properly acknowledged the anger that Robert had been feeling inside while on the surface he had been saying appreciative and nice things. I could also recognize how my need for Robert's gratitude had led to our colluding in niceness. Robert *was* appreciative, but he was also very envious of the fact I was fit and well, and going away on holiday, while he faced death.

In this case, I was able to get the help I needed from the ward sister. However, often colleagues are experiencing such similar feelings that they cannot contain them for each other. Such situations cry out for consultation to the staff group and/or management so that these feelings can be contained rather than acted out or projected. Just as I, in my role as chaplain, needed some space in order to regain my ability to think, so as not to go on holiday feeling excessively guilty or angry, so staff groups need space to understand what they are carrying psychologically as a result of the work they do. The hospice or other care institution may then be able to re-engage with its primary task. It is the ability to tolerate ambivalence that can restore integration and the capacity to think, or, in Kleinian terms, move a group from the paranoid-schizoid to the depressive position (see Chapter 1).

CONCLUSION

The tendency towards 'chronic niceness' is an aspect of the desire to be the perfect carer (see 'Guilt and reparation' in Chapter 12). This desire can create great stress for the individual carer, the group of carers and the patients. The recognition that one can perhaps be a *good-enough* carer for dying people or their families, without being perfect, can be very liberating. 'Out of the working through of the depressive position, there is further strengthening of the capacity to accept and tolerate conflict and ambivalence. One's work need no longer be experienced as perfect . . . because inevitable imperfection is no longer felt as bitter persecuting failure. Out of this mature resignation comes . . . true serenity, serenity which transcends imperfection by accepting it' (Jaques 1965: 246).

Dying people frequently create in their carers a desire to do everything possible to ensure that this part of the person's life is quality time. Striving for perfection may, however, cause considerable stress for all concerned. It may be more realistic to aim at being a nice-enough carer for that particular dying person. In this way, it may be possible to discover a strength which allows each to let go of the other in an appropriate and healthy way.

Chapter 11

Where angels fear to tread[1]
Idealism, despondency and inhibition of thought in hospital nursing

Anna Dartington

The organization of rudimentary care systems around the processes of birth, death, injury and disease is as old as human interest in the survival of the tribe. Indeed, it is the inevitable social necessity of nursing, whether professional or unpaid, which produces emotive and ambivalent attitudes to this work both in society and in nurses themselves. Nurses remind us, as doctors also do, of our potential vulnerability and dependency in illness, and of our mortality.

Society's contradictory attitudes to nursing, often referred to in the psychosocial literature as 'idealization' and 'denigration', are reflected in, for example, newspaper stories of selfless heroism among nurses, set against the reality of low pay and poor working conditions. The corresponding responses of the nurses themselves are flights of idealism, sometimes accompanied by illusions of omnipotence, and feelings of deep despondency.

Contemporary nursing is dogged by a negative expectation that nurses should not think. By thinking, I do not mean remembering whether Mr Jones is prescribed one sleeping tablet or two, but the processes of reflection about one's work, its efficacy and significance: registering what one observes of the patient's emotional state, the capacity to be informed by one's imagination and intuition, the opportunity to criticize constructively, and to influence the working environment. This is not to say that nurses do not think, but that it is an effort of will to make the space for reflection in a working life dominated by necessity, tradition and obedience. What is usually absent is the opportunity to ask the question 'Why?' of someone in authority, someone who is not surprised by the question, who is interested in the answer, and who can engage in a spirit of mutual enquiry.

Although the expectation not to think may have something to do with the denigration mentioned above, it is not simply a symptom of this. On the contrary, nurses are sometimes valued for a capacity to be passive at work, which itself is an example of the ideal of stoicism that pervades hospital culture. Nor can the expectation not to think be accounted for simply in terms of a doctor/nurse polarity of functions. It is true that many doctors are unreasonably expected to have a bright idea at a moment's notice, and are relentlessly pressurized to come

1 A fuller version of this chapter has been published in the journal *Winnicott Studies*, 7, Spring 1993 (pp. 21–41).

up with an answer when they have no answer to give. But nurses may also collude in their own unthinkingness, in the same way that doctors may sometimes collude with a pseudo-know-all-ness.

Those of us who think about unconscious processes in organizations are bound to look at the total organization for the origins of these collusive defensive patterns which are so wasteful of human resources. We could examine the nurses' collusiveness with a non-thinking expectation in terms of the conventional role of women at work. Gender issues can hardly be irrelevant in the consideration of a largely female workforce. However, while general nurses' working so consistently in a 'maternal' role is highly significant in many ways, it has been my experience that nurses feel themselves to be oppressed not by men *per se*, but by social systems. In the context of this book, what is of most interest are institutional and societal dynamics that can be observed that will explain the necessity for an unthinking workforce operating at the interface between the patient and the institution.

THE IMPACT OF THE HOSPITAL CULTURE

Most people who have spent time in hospital either as patients or employees will remember the initial impact of the sights, sounds and smells that assailed them as they left the world behind. The memories linger because they are associated with fear. It is perfectly understandable to feel afraid when we put our bodies, and possibly our lives, in the hands of strangers. It is more difficult for staff to admit their own fear: that because strangers are prepared to trust them, the burden of responsibility and its attendant dependency might prove overwhelming.

I will now discuss the impact of entering a new institution from two different vantage points: one as a nurse student myself, the other, twenty-five years later, as a group-work consultant to nurse tutors. Both hospitals were large London teaching hospitals.

Some personal recollections

I joined the nurse training school soon after my eighteenth birthday. Most of the students were middle-class, and the majority had chosen nurse training as an alternative to university. Several were devout Christians and viewed nursing as a form of religious vocation. The culture was largely serious, committed and inhibited.

When talking to my friends who were at university, I realized that, by comparison, we were given little opportunity to question our teachers. Certainly we had not encountered seminars; although the nursing school seemed to pride itself on recruiting intelligent young women, it then seemed incurious about what was going on in their minds. Our teachers were pleasant but consistently distant; the expectation was that we would not be curious either. We asked questions about facts, of course, but we lacked both the innocence and the ruthlessness of any free-thinking three-year-old. We remained silent, possibly experiencing the teachers as rather fragile.

We had yet to learn that we too were fragile, afraid of all the horrors the hospital might contain. The memory of my first visit to a surgical ward remains particularly vivid.

I stood at the end of the bed of someone with tubes coming out of every orifice. He was gripping so tightly to the bedframe that his knuckles were white. I felt giddy and faint. In my imagination the man was being tortured, a thought so terrible I could not even voice it to my friend, who was asking sensible questions about the temperature chart. Later, as I sat recovering in the cool corridor, I felt foolish. It occurred to me that the new patients arriving on the ward might have similar waking nightmares, and like me feel ashamed of themselves.

From this experience I learned something of how hospitals are bursting with intense primitive anxieties about the potential sadistic abuse of the power staff have over patients, and that everybody has these dreadful thoughts but nobody ever speaks about them. Staff and patients alike are expected to exercise stoicism and repression. Once I started working on the wards, I found these fears lessened. I could chat to the man who had clung to his bed. He might ask me the results of a football match, or I might change his urine bag. In this way I was reassured, and, in the course of these ordinary and practical exchanges, so was he. It is often easier to be an active participant than to be an observer in hospitals.

My other abiding recollection is from my second week on my first ward, a male medical ward.

Most of the patients were much older than the new nurses, except for a boy aged nineteen who was pleased to have other young people to talk to. For me, he represented a link with the more light-hearted world of my friends outside the hospital. In those days, the wards were well staffed, and there were opportunities to talk to patients. David had been admitted for tests. He looked healthy and handsome, and quite out of place among the elderly coronary patients.

One afternoon, my friend Kate, a fellow student, erupted into the ward kitchen. She had overheard a conversation between the consultant and the ward sister – they had been discussing David's imminent death. I wanted to disbelieve her, but I knew from Kate's expression that it must be true. 'It's leukaemia,' she added. I continued with my duties, leaden with anger. I avoided David, but at one point he called me over. 'I was wondering,' he said, 'if you'd like this box of chocolates. I've got too many.' Without thinking, I remonstrated, 'No, you keep them. You'll enjoy them when you're better.' David looked at me with exasperation and pity. 'I'd like you to have them,' was all he said. I accepted the chocolates, and he accepted the weight of my apology. David died three days later.

Three years earlier, Isabel Menzies had written in her now famous study of the nursing service of a teaching hospital: 'There is no individual supervision of student nurses, and no small-group teaching event concerned specifically to help student nurses work over the impact of their first essays in nursing practice and handle more effectively their relations with patients and their own emotional reactions' (1960: 61).

I needed to understand why David had died. The ward sister made time to speak to me in broad terms about leukaemia. 'You'll get used to it,' she said. It was meant kindly, but what did it actually mean? Her invitation was to the world of experience, of greater objectivity; but without personal supervision or the opportunity for mental digestion, it could only become an invitation to avoidance and denial. In such well-intentioned and innocent ways the system perpetuates its defensive organization against anxiety.

The student nurse group project

Twenty-five years later, I was approached by a nurse tutor at a large teaching hospital, who had noticed her students were increasingly raising issues to do with their feelings about patients. She asked me to join her in a project in which weekly discussion groups would be offered to new nurses during the time of their early ward placements. Each group would be led by a tutor, and I would act as group consultant to the tutors themselves.

The students seemed a confident group; the mixture of sexes, styles, races and accents gave the overall impression of a comprehensive school sixth form. The tutors were welcoming and informal. They appreciated their colleague's initiative because they thought it would support the students. The notion of support was always prominent, and the groups came to be known as support groups, although I preferred to think of them as exploratory meetings about work. The danger, as I saw it, was that support in the form of reassurance would replace exploration, but I was regarded as fussy and somewhat truculent for taking this stance.

The initial student take-up of the groups was about 60 per cent. The older the students, the more interested they seemed to be. One hypothesis was that the younger students were still struggling with adolescent identity issues, and that this, together with the pressures of assuming a new role in a new institution, would have represented a considerable onslaught on their psychic defence mechanisms. Any additional emotional exposure in a group could have been anticipated as intolerable.

The project started with four groups of six or seven members. The themes of the groups, as highlighted by the students, are broadly categorized below:

- Potential and actual abuse of power by hospital staff; for example, forcing unwanted treatments on a terminally ill patient.
- Dealing with patients who exercise independent thought; for example, the Roman Catholic patient who objected to being in a gynaecology ward where patients were being admitted for termination of pregnancy (the administration refused to move her).

- Coping with one's own response to tragedy while maintaining a professional role; for example, with regard to a teenager, brain-damaged in an accident, who will never regain consciousness.
- Shame associated with mistakes; for example, forgetting a patient had requested her son should be called before her death.

The tutors were quite shaken by some of the incidents and dilemmas described in the groups; the discussions forced them to think about something they had forgotten or repressed. They became preoccupied with the discrepancy between what was taught in the nursing school and the reality of busy ward practice. The tutors felt the sense of disillusion and helplessness that the nurses experienced. Sometimes they felt it necessary to give talks on topics such as how to care for the dying, to inject not only concern for standards but also some hope into the situation. Providing a forum for the students to think had, it seemed, pushed the pain of helplessness and of failed idealism upwards into another part of the nursing system. The tutor group became restless, and their attendance at the supervision group became erratic.

Some of the tutors were not particularly willing to acknowledge the degree of mental pain among the students, and at times I felt quite irritated by their unwillingness to try a more interpretative approach, which perhaps could help the students to feel contained and understood. For a while there was a wish to anaesthetize me. Did I represent an anarchic threat to the hospital? Although this seemed a mad and grandiose idea, I felt I was being experienced as pushing them into something dangerous. Gradually it became clearer. The consequence of an accurate and helpful interpretation is likely to be increased intimacy and dependency. It was as if, once the students were aware of the extent of their distress at work, they would lay their burdens at the door of the tutors: 'You have helped me to recognize this. Now what are you going to do about it?' The tutors imagined they would be held responsible. This seemed to mirror the students' fear of direct personal contact with patients and attachments that would involve impossible demands.

What I, the students and the tutors were all experiencing at first hand were the unconscious assumptions of the hospital system, which were that attachment should be avoided for fear of being overwhelmed by emotional demands that may threaten competence; and that dependency on colleagues and superiors should be avoided. One should manage stoically, not make demands of others, and be prepared to stifle one's individual response.

If, for a moment, we consider the institution as the patient, it is as if emotional dependency is experienced as the most dangerous and contagious of diseases. Everyone is under suspicion as a potential carrier, and an epidemic of possibly fatal proportions is likely to break out at any moment. The only known method of prevention is stoicism, which is administered by example and washed down with false reassurance. Since the patient/consumer is already seriously infected, by virtue of their institutionalized role, he or she must be kept at a courteous but safe distance.

We all know of small-scale health care institutions, such as hospices, where appropriate dependency in patients and staff is acknowledged with remarkably favourable consequences. Unfortunately, most nurses are trained in large hospitals, where institutional processes lead to a kind of madness in which dependency is simultaneously encouraged and punished.

MOTIVATION, FRUSTRATION AND SATISFACTION

Despite the stress and distress of institutional life, nurses continue to find pleasure and satisfaction in nursing. Job satisfaction is notoriously difficult to define. The reasons why we choose the work we do and the elements that provide satisfaction are largely unconscious and related to complicated emotional needs (Main 1968; see also Chapter 12). However, it is possible to make the general observation that many people are drawn to the caring professions because they have a need to put something right. This need to make reparation may be only partially conscious. It arises from guilt or concern, and its aim is to heal emotional wounds: one's own, and those of the damaged figures of one's internal world. It can be expressed in a socially acceptable way by helping others. While reparative wishes are healthy (one might even say that to heal ourselves while promoting healing in others is a good psychic economy), problems can arise when there is a compulsive quality. We see this, for example, in the driven motivation of the do-gooder who appears self-satisfied, out of touch with his or her own vulnerability, seeing problems only in others, and determined to 'save' people with or without their consent.

Whatever it is that engenders the wish to nurse, job satisfaction will depend not simply on the presence of a patient, but also on the presence of a patient who needs a nurse.

I was asked by a trainee psychotherapist, Jane, to supervise her on a difficult piece of work. She had been asked to offer some support to a group of qualified nurses who worked on a liver unit. The nurses had been suffering from stress. Some had complained of it openly, others had taken long periods of sick-leave for apparently minor illnesses, one had resigned, and still others had asked to be transferred to another unit. The hospital administrator had been alerted, and some money was made available to provide staff support. A series of whole-day meetings was organized at which Jane intended to offer the nurses an opportunity to relax and to talk about their stress. Out of a total of twelve, only five wished to attend.

Jane had planned to start with some brief relaxation exercises, but the nurses were unable to relax. One lay on the floor like an ironing board, another said surely it was all pointless, and two others said they were afraid of what might 'come out'. Jane then encouraged them to sit in a group and just talk.

They talked, not surprisingly, about the number of deaths on the ward, the liver transplant patients whom they lost after days or weeks of painstaking intensive care, the distress of the relatives, their own exhaustion. The most difficult thing to talk about was the resentment and even hatred they sometimes felt towards the patients who seemed to thwart their every effort and care. These were the alcohol- and drug-dependent patients who returned again and again to be repaired, with the single intention, or so it seemed to these nurses, to continue to connive at their own death.

These nurses may or may not have been sophisticated in their knowledge of the complexity and intransigence of addiction. Our intellectual knowledge often makes very little difference when we are faced, day after day, with the hopelessness of persistent self-destructiveness in others. The nurses felt deprived of the opportunity to make reparation; they could not experience any gratification or significance in their work because they could not help the patients to get better. Jane could at least help these nurses to recognize the basis of their stress. Acknowledging and understanding their hatred of their patients for (as it seemed to them) refusing to get better could help them not to retaliate by ignoring the patients' distress.

In this sort of extreme frustration of work satisfaction, workers need opportunities to mobilize appropriate defences against pain and anger. For this to happen there has to be someone, most helpfully a senior colleague, available to share the burden. In a highly stressed and increasingly petrified system, it may be necessary for an outsider to be called in to ask the question 'Why?', and to be a catalyst and container for thinking.

If there are no such opportunities, and appropriate defences are consistently blocked over time, there are two types of response. One is the breakdown/breakout solution. This would include the development of psychosomatic symptoms, the avoidance of work, long periods of sick-leave, depression and, ultimately, resignation. Such things get noticed, of course, as happened in the liver unit, when it led rather late in the day to an offer of support to the remaining staff.

A second solution is the evolution of pathological psychic defences in the worker. This does not get noticed in the same way because it develops gradually. Sometimes people erect a shell around themselves which serves to deflect and anaesthetize emotion. This is largely an unconscious development. If such a shell becomes a permanent feature of the personality, it is at great cost to the individual, who can no longer be fully responsive to his or her emotional environment. The resulting detachment is dangerous to clients, patients or colleagues, who will sense the potential cruelty inherent in the indifference.

Appropriate defences are those which are mobilized in the recognition that a situation is painful or downright unbearable. They involve attempts to protect oneself from stress in order that the work-task may be preserved. Pathological defences are those which are mobilized in order to deny reality, to allow a really mad or really unbearable situation to continue as if it were perfectly acceptable, when in fact it needs to be challenged in order to preserve both the workers and the work-task. The soft end of pathological defence is stoicism; the hard end is manic denial, a psychotic process that attempts to obliterate despair by manufacturing excitement. In manic states of mind people are oblivious to both pain and danger.

MATERNAL TRANSFERENCE AND COUNTERTRANSFERENCE

Dr D. Winnicott, who helped mothers to feel that being merely 'good enough' was all right, extended his ideas to those in the helping professions. He listed many reasons why an ordinary mother may at times hate her ordinary baby, among them that the baby 'is ruthless, treats her as scum, an unpaid servant, a slave' (Winnicott 1947: 201). In caring for disturbed, frightened, angry, dependent patients, nurses may experience similar feelings. The constant demands to be in attendance and available can give rise to much resentment and self-denigration, in the way that mothers at home with children complain they feel relegated to a low-status activity. They complain about this despite the fact that they love their children and want to be with them. Availability seems to be associated with denigration. We could say that the good-enough patient is one who provides enough job satisfaction to enable the nurse to continue with her work despite its unpleasant aspects. The good-enough patient senses that the nurse needs to be needed, as well as needing some protection against excessive emotional demands.

There are inevitably intimate moments between nurse and patient. The nurse knows she will be remembered. She knows, too, that very strong feelings will temporarily be transferred on to her. She knows that when patients wake up after major, life-risking surgery and see her face, they may experience her as an angel, not because they are hallucinating, but because she is associated at that moment with the beauty of being alive. At such a moment, each will experience the individuality of the other. Yet the hospital culture, as we have seen, does not encourage the nurses to be moved by their experiences; attachment is felt as a threat to the system. Nurses therefore often keep these moments of intense job satisfaction to themselves. This may also be because the patient unconsciously conveys embarrassment to the nurse, embarrassment about being an adult but feeling like an infant. The nurse 'holds' the embarrassment for the patient, and does not speak about it. It is as if a pact is struck between them: the fulfilment of the nurse's reparative and maternal instincts, and the patient's appropriate dependency needs.

CONCLUSION

Many nurses have told me that when they step outside the training environment of the hospital to take up various nursing roles in the community, they experience an acute fear of their new professional autonomy, and also of the autonomy of their patients. Community nurses become retrospectively aware of their previous protected and even infantilized status as professionals who were not expected to think for themselves, nor to take initiatives while working in hospitals. If the institutionalization has not become chronic, and if the new working environment is sufficiently tolerant of individuality, nurses will in time re-discover their curiosity and capacity for thought.

Most of us have an unthinking area of our work where we operate on assumptions. These are subjectively experienced as innocuous habits, but when challenged, despite the fact that no one is able to articulate the reason for them, the rules persist. These no-go areas of institutional life are usually those in which the most anxiety is generated. These areas of anxiety in society's humane institutions can be identified with ways of managing the containment of suffering, death and fear, in the clients and the workers. The intense emotions aroused are felt to threaten not only efficiency, but also the fabric of the institution itself. In a vast and perhaps literally unmanageable organization like a national health service, it seems to be the fate of those who work on the staff/client boundary to carry and attempt to contain this anxiety so that the rest of the organization can experience an emotion-free zone in which to operate. In order to maintain this, these frontline workers must be silenced, anaesthetized, infantilized or otherwise rendered powerless. Student nurses in particular often feel disempowered by infantilization; because they feel afraid but must pretend to be strong for the patients, they experience themselves as merely playing nurses. They are frequently asked, 'Did you always want to be a nurse?', the implication being that this career choice did, or should, stem from some sweet and innocent notion.

Nurses are reacting to the imposed non-thinking caricature by an intense period of professionalization, in which increased opportunities to take degrees and postgraduate diplomas are encouraged. Under Project 2000, student nurses will spend increasing amounts of time away from the wards in institutes of higher education. Trained nurses will care for the patients, supported by unqualified workers called health care assistants. The danger is that the health care assistants will come to bear the brunt instead of the student nurses.

One can only hope that improved morale in one socially essential profession may have positive consequences in wider systems of social care management, causing old assumptions to be questioned. If, on the other hand, we continue to behave as if emotionality in the workplace is best managed by denial, splitting and projection, then we will continue to inhibit the functioning of society's humane institutions, and continue to squander the potential thoughtfulness of those who work within them.

Chapter 12

The self-assigned impossible task

Vega Zagier Roberts

The preceding chapters in this part of the book have described how the nature of the work in various settings across the helping services affects the workers, giving rise to collective or institutional defences, which in turn determine organizational structures and practices. This chapter looks at what the workers bring to the work, their needs and inner conflicts, and how these make them particularly vulnerable to getting caught up in the institutional defences arising from shared anxieties.

CHOOSING A HELPING PROFESSION

The choices we make regarding which profession to train for, which client group we will work with, and in what kind of setting, are all profoundly influenced by our need to come to terms with unresolved issues from our past, as the following example shows:

> Marian was the eldest of six children who were abandoned by their alcoholic father when she was thirteen. Their mother, who had to go out to work, looked to Marian to care for the younger children, often repeating, 'I don't know what I would do without her.' This left Marian little time for her studies, and she was the only one of the six who did not go to university. She stayed at home until the youngest child passed his 'O' levels. At the age of thirty, Marian found her first paid job, as an assistant in an inner-city nursery school for disabled children. Here she maintained a familiar source of self-esteem, dedicating herself to helping young children get the best possible start in life. Like her mother, the teachers at the school often wondered how they would manage without Marian. At a less conscious level, Marian was working through issues related to a view of herself as handicapped by her lack of education, and by the social isolation of her adolescent years.
>
> Violet, one of Marian's younger sisters, who had been only four when her father left, became a family therapist. While she was consciously motivated by a powerful desire to keep families together, it was a constant battle for her to control her rage at parents when they behaved in ways she considered harmful to their children. This made it difficult for Violet to work effectively,

until personal therapy enabled her to disentangle her own experiences from those of her clients.

A brother, Harry, had been an unruly child, and was often told he was making life even more difficult for his mother. Harry became a teacher, and later headmaster of a boys' boarding school. He was a stern disciplinarian, and his internal image of his school was of a place where unruly boys were sent to be kept under control, and to relieve parents of a burden they could not cope with.

IDEALS AND DEFENCES

Many of the conscious choices made by helping professionals are based on idealism. However, ideals also have unconscious determinants, and these can contribute to defensive institutional processes.

Fairlea Manor was a private psychiatric hospital, famous since the 1940s for its pioneering work in applying psychoanalytic methods to the treatment of severely disturbed and previously untreatable patients. It retained a reputation as a place of hope for people who no longer responded to more conventional psychiatric intervention. At a case conference, Veronica, a fairly new psychotherapist, presented Dan, a young man with whom she had been working for fifteen months. As was usual at Fairlea Manor, all his medication had been stopped when he arrived, and for many months he had been very agitated and often assaultive. Now, although he was still too disturbed to leave the ward unescorted, his symptoms were subsiding. Veronica thought that together they were coming to understand his illness, and was hopeful that Dan would be one of the rare and highly-prized patients who recovered without medication. However, the ward staff were pressing for a decision to re-institute medication to hasten Dan's improvement. This was agreed at the case conference, over Veronica's strenuous objections. She was angry and bitterly disappointed, experiencing the decision and the nursing staff's focus on promoting 'normal' behaviour as an attack on her work. In supervision, she described feeling drugged and sleepy during sessions with Dan, as if she too were being medicated. For their part, the staff regarded Veronica as dangerously out of touch with the reality of Dan's time-limited health insurance and his lack of real progress towards being able to live outside the hospital.

This was not uncommon. Acrimonious arguments between therapists and the nursing staff were a regular feature of case conferences, and there was little dialogue between the two groups. Many of the therapists had come to Fairlea Manor because they believed passionately in the value of psychoanalysis and this was one of the few hospitals where it was still a core treatment. They disregarded the external reality that the environment had changed dramatically since the 1940s, when the absence of alternative treatments meant that patients could remain at Fairlea Manor indefinitely if necessary. Now the insurance companies paying for treatment demanded that hospitals demonstrate they were keeping patients' length of stay to a minimum.

This disregard, even hatred, of external reality is typical of the basic assumption mode of group functioning, where the task pursued by a group is more the meeting of members' internal needs than the work-task for which it was called into being (see Chapter 2). It is associated with an absence of scientific curiosity about the group's effectiveness, an inability to think, learn from experience, or adapt to change, and is most likely to dominate when there is anxiety about survival (Bion 1961). The therapists were under threat in a number of ways, not least from the nature of their work with such deeply disturbed people, its assaults on their safety and sanity, its unrelenting demands and its often disappointing results. In addition, their survival as specialized professionals was threatened by the decline of psychoanalytic psychotherapy as a psychiatric intervention. Finally, many were still in analytic treatment themselves as part of their training, and deeply needed to believe in its efficacy.

In response, they set themselves an impossible task: to prove they could cure any mental illness, however serious, with psychotherapy. No one defined their task to them in this way; it was self-imposed, unarticulated, powerful and very persecutory, since, thus defined, it was unachievable. They had to find most of their satisfaction in their endurance, and in their shared disdain for those seeking merely external, superficial improvement in the patients. They split off any doubts they had and projected them into the ward staff. They also split off their rage at patients who refused to get better, and blamed the hospital for not providing the resources – especially unlimited time – which would enable them to achieve their self-assigned impossible task.

Meanwhile, they unconsciously entrusted to the ward staff the task of 'exporting' sufficient numbers of patients who had recovered enough to return to the community, the task on which the hospital's economic survival depended. To the outside world, Fairlea Manor presented a united front, defending its costly and time-consuming approach to treatment from all criticism. In this way, it provided the therapists with a place where they could continue to do the work they needed to do in order to meet their unconscious needs.

TASK PERFORMANCE AND UNCONSCIOUS NEEDS

To the extent that people are drawn to work in a particular setting because it offers opportunities to work through their own unresolved issues, these settings may well attract staff with similar internal needs and a similar propensity to fit with certain kinds of defences. Bion (1961) refers to this phenomenon as *valency*, and its part in determining one's choice of profession has been discussed in Chapter 2. This gives rise to collective defences against the anxieties stirred up by the work which can seriously impede the task performance.

Swallow House was a social services residential unit for children who had been removed from their families for their own safety. Its task was to prepare for the children's return to family life, either with their original parents or with adoptive parents. Although discharge planning was ostensibly central from the first day, in practice the staff focused their efforts on providing as secure and home-like a place as possible. This in itself was positive, but difficulties arose when it was time for a child to move on; the staff could never see any parents as being good enough or ready enough to take the child on. It was as if only Swallow House staff could meet the children's needs, and almost every departure was traumatic.

When the manager tried to help the staff with this in supervision, his efforts were violently resisted. Staff were very suspicious and distrustful of the manager, casting him in the role of a 'bad parent' who was too preoccupied with his own concerns to have the children's best interests at heart.

The self-assigned impossible task in this case seemed to be to provide the children with the (ideal) parenting they had never had: impossible in any case, but actually anti-task in an institution intended for children in transition. This very young staff's unresolved issues – several had themselves been in care, or came from broken homes – both about themselves as children, and also as potential parents, led on the one hand to unrealistic expectations of the parents to whose care the children were discharged, and on the other to attacks on their manager for failing to be an ideal parent to them. This made it difficult to make use of the 'parenting' he did offer; supervision sessions were often cancelled, and his authority constantly challenged.

The staff over-identified with the children, partly as a result of the key-working system, which meant that children were assigned to particular staff members who took primary responsibility for them. This system, originally designed to reduce the psychological trauma of institutional care, put a dys-functionally strong boundary around the child/key-worker pair, a boundary which supported the as-if task at the expense of the task which needed to be done (see Chapter 3). The special and exclusive intimacy of key-working, consciously intended to meet the needs of the children, was also a product of the need for intimacy in the staff.

PROFESSIONAL IDEALISM AND GROUP IDENTITY

At both Fairlea Manor and Swallow House there were difficulties in the export process – discharging the patients or children. This had to do with a collective sense of everything good and helpful being inside the organization, and of the outside world as harmful and dangerous. In both institutions, group identity was based on being a superior alternative to another form of care. At Fairlea Manor, this was explicit: it set out to be an alternative to 'revolving-door' hospitals offering short-term treatments aimed at symptom control. At Swallow House, it was less conscious: key-workers strove to be better parents to the children than those in whose care they had previously been, whom the staff blamed for the children's situation.

A group's identity is linked to its definition of its primary task – its reason for existing (Rice 1963; see also Chapter 3). To some extent this always includes a dimension of being an alternative to some other group. 'We make shoes' is a start, but we identify ourselves with a particular group making particular shoes, more stylish, cheaper or more comfortable than others. In organizations caring for people, identity and task are often linked with ideals and ideology. For example, Chapter 8 described models of care based on a 'warehousing' ideology, where patients are treated as creatures with purely physical needs, and an 'horticultural' ideology, where patients are treated as individuals with unfulfilled potential which needs to be developed (Miller and Gwynne 1972). Workers using this second approach tended to be so persuaded of its superiority that it was difficult for them to recognize how it, too, failed some of their clients.

Since the personal meaning of the work tends to be vested in the ideals underlying the choice of working methods, it can be very anxiety-provoking to question them. Instead of space to reflect on what is most appropriate for whom, there is often polarization around 'right or wrong', as illustrated in the following example:

> At the Tappenly Drug Dependency Unit (previously described in Chapter 3), the growing waiting list became such a pressing problem that the service managers asked the staff to revise their methods, so as to reduce the list as quickly as possible. The team saw this as devaluing their good work, and united to defend long-term counselling, without showing any interest in comparing outcomes for clients receiving different amounts or types of counselling. The argument, which was bitter, remained anecdotal and highly personalized.
>
> As reforms in the health service threatened to make evaluation a requirement, staff predicted that this would be based purely on numbers, without any regard for clients' welfare; yet they were unable to suggest any alternative criteria when they were invited to draft their own. The introduction of a computer to collect data was perceived as a way for managers to spy on and control the unit, rather than as an opportunity to gather the information they needed for service-planning. When a researcher offered to train them how to use the computerized data to answer their own questions about how clients were using their service, they were so suspicious that he was an auditor that they were unable to formulate questions that might have enabled them to use their limited resources more judiciously.

Many teams and organizations are set up as alternatives to other, more traditional ones, often by someone disaffected by personal or professional experience of other settings. However, identity based on being an alternative, superior by some ethical or humanitarian criterion, tends to stifle internal debate. Doubts and disagreement are projected, fuelling intergroup conflict, but within the group everyone must support the ideology. Any questioning from within the group is treated as a betrayal of the shared vision. In the Tappenly team, for example, some doctors and nurses ventured to express alarm about the risks to health and life for people on the waiting list, and suggested diffidently that it might be better to take some staff time away from counselling to assess the physical state of the waiting clients. The panic and anger which greeted this 'selling out' to the opposition quickly led to the proposal being dropped. Similarly, some Fairlea Manor therapists would from time to time question whether the use of medication or more emphasis on symptomatic improvement necessarily undermined the psychoanalytic work, but the prevailing anxiety about 'becoming just like anywhere else' made this feel too dangerous to talk about. Thus, internal differences were constantly stifled (see Chapter 16).

The self-assigned impossible task becomes the 'glue' which holds these alternative organizations together. It serves as a defence against the difficulty and stress of working at the possible. Problems inevitably arise when the alternative approach proves limited. Working with chronic schizophrenics or abused children or heroin addicts is intrinsically difficult, and success is never as great as one hopes. The alternative approach is based on a hypothesis that by changing certain conditions far more success will be achieved. When outcomes fail to support this belief, disappointment and anxiety set in. Failure is experienced as a punishment or envious attack for having dared to set themselves up as a superior alternative. This serves to reduce guilt and anxiety about the institution's shortcomings, but at the cost of leaving staff feeling under attack, helpless, abused and unable to learn from their experience.

GUILT AND REPARATION

From a psychoanalytic point of view, it is the drive to effect reparation, partly conscious, but largely unconscious, that is the fundamental impetus to all creative, productive and caring activities. This drive has its roots in our own experiences with our earliest caretakers – let us say with mother. In the earliest months of life, the infant splits his perception of mother into good and bad: the good, nurturing mother he loves, and the bad, depriving mother he hates and attacks. With maturation comes the awareness that mother is a single person for whom the baby feels both love and hate. With this realization come remorse, concern and guilt for the damage his greed and aggression have caused and will cause her. From this stems the drive to reparation: to atone, to protect, and also to express gratitude for the good care received. Normally, the child's reparative activities will keep his anxieties at bay. When guilt is too strong, however, reparative activity will be inhibited. Instead, the infant – and the adult in whom these early conflicts are revived – retreats to the earlier, more primitive mental activity of splitting objects into good and bad, which can be unambivalently and separately loved and hated.

Usually, by repeatedly discovering that mother and, later, others survive his attacks, the infant comes to trust that his love predominates over his hate, and that his reparative activities are successful. This lessens his fears of persecution and retaliation by the bad mother he has attacked. But when external reality fails to disprove the child's anxieties, for example, if the mother dies, or withdraws, or retaliates, then depressive anxieties may become too great to bear. The individual then gives up his unsuccessful reparative activities, resorting instead to more primitive paranoid, manic and obsessional defences.

Paranoid defences involve denial and projection of aggression so that it is experienced as coming from outside oneself in the form of persecutors. Manic defences are directed at denying that damage has occurred, and involve omnipotent fantasies about magical repair. Unlike genuine reparation, which requires the ability to face that damage has been done and cannot be undone, manic reparation must be total, so that no anxiety, grief or guilt need be experienced. Manic reparation tends to be impractical and ineffective; when it fails, individuals may use obsessional defences, the ritualistic repetition of certain acts, as a further effort to control and master anxiety, especially about their aggressive impulses (Klein 1959; Segal 1986). Those in the helping professions inevitably and repeatedly encounter failure in their work with damaged and deprived clients. If this arouses intolerable guilt and anxiety, they, like the infant, may retreat to these primitive defences in order to maintain precarious self-esteem, and to defend themselves against the retaliation anticipated for failing to heal.

Two features distinguish caring work from most other work. The first is that the reparative activities are carried out in direct relation to other human beings. This means the job situation often very closely resembles early-life situations that the worker may still unconsciously be dealing with, and which drew him or her to this particular line of work. The second feature is that the worker's self is felt to be the major tool for producing benefit for the client. The helping professions are often regarded both by their members and by their clients as vocations requiring special qualities. Skills and 'technology' – the syllabus, the treatment programme – may be used defensively to ward off anxiety about personal adequacy for the task. Often, however, they are put in second place, with primacy being given to personal attributes as the instrument of change. By offering themselves as such instruments, workers unconsciously hope to confirm that they have sufficient internal goodness to repair damage in others. This is a source both of individual and organizational ideals, but also of much anxiety.

Haply Lodge was founded as an alternative to statutory provision for homeless mentally ill people. The staff strove to abolish all distinctions between themselves and those they served, living with them in a community where the tasks of daily living, such as cooking and cleaning, were shared by all. There were no rotas or shifts: everyone lived full-time at Haply Lodge, with minimal personal privacy and no demarcations between on-duty and off-duty periods. The vision, to which staff felt deeply committed, was that by doing away with any barriers between themselves and their clients, such as they believed had contributed to the reluctance of many homeless people to use statutory services, they would bring about healing through love and mutual respect. That the absence of limits on what they offered generally led to their own physical and emotional breakdown within a year or two was built into the culture of the organization. Despite the fact that this burn-out was expected, it was usually accompanied by disappointment, anger and guilt in both management and workers.

When the consultant explored with staff how they had come to work at Haply Lodge, it turned out that nearly all of them had painful personal experiences of homelessness – actual or psychological – and had no meaningful home to go to, were they to work more conventional hours. They found the intimacy and sense of belonging at Haply Lodge deeply rewarding, at least at first, and were shocked when their clients drifted back to 'life under the bridges' as they always had.

Although the impossible task here was explicit rather than self-assigned, this example highlights unconscious processes which operate across the caring services. Staff at Haply Lodge were stressed both by the similarities and by the differences between themselves and their clients, and both played a part in their choosing this particular work. They felt guilt for being better off than their clients, having more education, skills and material resources, and therefore more choices about where and how to live; guilt, too, for being glad this was the case. By obliterating these differences, the staff unconsciously sought both to reduce their guilt and to find the loving home missing in their own lives.

The aim of helping or healing through one's love and goodness is two-pronged. Success is deeply validating, strengthening the capacity to act constructively. But failure, or even limited success, is felt to demonstrate inner deficiency, and is intolerable. At Haply Lodge, the burn-out served to assuage guilt about lacking sufficient goodness to bring about the intended result. It was also the only way staff members could get away, since their investment in the organization's ideals precluded their actively choosing to leave at an earlier stage.

CONCLUSION: MANAGING ANXIETY IN THE HELPING SERVICES

To understand, and therefore to be able to help, another person requires a capacity for empathy: to stand momentarily in the other's shoes and experience their pain, using what one has learned as a guide as to how best to respond. However, the close resemblance between workers' own most painful and conflicted past experiences and their experiences at work constantly threatens this capacity. In some institutions, the dominant defence is to accentuate differences: 'they' (the clients) are the sick or mad or needy ones; 'we' (the staff) are the well, sane, strong, helping ones. The work in this case will be structured to support the distance between staff and clients, using rigid time-tables, programmes and hierarchies. In other institutions, particularly those dominated by a self-assigned impossible task, the dominant defence is more likely to be to deny differences, to stand so much in the others' shoes – identifying with the clients as victims – as to be overwhelmed by their pain and despair. In both cases, there is a failure to manage the client–worker boundary in ways that support task performance (see Chapter 3).

It is therefore of the greatest importance for helping professionals to have some insight into their reasons for choosing the particular kind of work or setting in which they find themselves, and awareness of their specific blind spots: their valency for certain kinds of defences, and their vulnerability to particular kinds of projective identification. It can be helpful to explore this with colleagues, in the process gathering clues as to the source of some collective defences. Personal therapy can also be of assistance, as in the case of Violet (described at the beginning of this chapter), to disentangle one's past from the present, and to find alternative ways of resolving unconscious conflicts, rather than needing to do this entirely through one's work.

By recognizing consciously the internal task definition they are working to, staff groups can become more aware of the associated anxieties and defences. A more realistic task definition can then be formulated, one that is meaningful but also possible, and which relates to the overall task of the wider institution. Realistic task definition, by making some success possible, increases the capacity to tolerate depressive anxieties. Managers can take a lead in this, provided they maintain their position at the boundary of the system and are not too caught up in the defensive processes themselves (see Chapter 3).

Coherent thought and the capacity for problem-solving are possible only when depressive anxieties – and hence reality – can be tolerated. When paranoid anxieties prevail, thought and memory, which link reality to consciousness, have to be attacked and eliminated, and there is no curiosity about causation. In this case, problems cannot even be stated, let alone solved (Bion 1967). In the depressive position, omnipotent fantasy, obsessional ritual and paranoid blaming can give way to thinking: one can seek to know, to learn from experience and to solve problems. Reparative activities can then become more realistic and practical, allowing workers more solid satisfaction from their very difficult work.

Part III

Institutions in crisis

INTRODUCTION

The sense of crisis is widespread in society today, and nowhere more so than in the human services, where each day brings news of further cuts, restructuring and other changes which threaten our personal security and professional practice. Inevitably, these cause great stress. Stress can also arise in response to internal crisis; for example, when there is intractable conflict among staff, or when high levels of absenteeism increase the pressure on those who remain. This part of the book considers various kinds of crises across a wide range of helping organizations, and how these are managed.

Chapter 13 examines the links between very rapid organizational change and the increase in personal and interpersonal stress experienced by individuals. It is important to distinguish staff stress produced by organizational factors from that produced by personal ones if the underlying difficulties are to be tackled adequately. Chapter 14 develops one of the themes of the preceding chapter, illustrating how organizational difficulties can be attributed to a particular member of staff, who is then scapegoated and even expelled from the institution in the mistaken belief that this provides a 'solution' for the problem. In other situations, different members of the organization 'hold' or express different aspects of fundamental organizational dilemmas, thereby colluding unconsciously in a collective avoidance of difficult issues which need to be addressed.

Chapter 15 illustrates the cost to a particular organization of not facing the uncertainties confronting it. By putting into words what had previously been too dreadful to talk about, and therefore impossible to think about, its members were enabled to begin planning for their own and their clients' futures more effectively. Chapter 16 gives further examples of situations where crisis is dealt with by denial, focusing particularly on the cost of avoiding facing differences within groups. In this case, no member of the group can be empowered to act or speak on its behalf, and opportunities for negotiation may be lost.

One of the commonest requests for help in times of crisis from teams in the helping services is for a staff support or sensitivity group. The last chapter in this part is therefore devoted to exploring some of the factors that can undermine or contribute to the usefulness of such groups. Many of the issues raised, however, also apply to other kinds of consultative interventions. As already illustrated in earlier chapters, requests for help can be invitations to displace problems or avoid them altogether. One needs to be constantly alert to the possible covert and unconscious aims of the request, so as not to get caught up in an unhelpful collusive process.

Chapter 13

Institutional chaos and personal stress

Jon Stokes

In this chapter I wish to make some links between the increasingly uncertain and sometimes chaotic nature of life in present-day organizations and the personal experiences of stress felt by those working in them. Essentially my argument is that in order to understand many apparently personal experiences of stress, it is important to place these in their organizational context of uncertainty about the future, and a related confusion about the organization's primary purpose or mission.

THE ORGANIZATION-IN-THE-MIND

The conceptual framework I wish to use is based on the idea of an organization or institution 'in the mind'. This term was first introduced by Pierre Turquet in the 1960s when he was working at the Tavistock on group relations with applications to organizations (see also Armstrong 1991). It refers to the idea of the institution that each individual member carries in his or her mind. Members from different parts of the same organization may have different pictures and these may be in contradiction to one another. Although often partly unconscious, these pictures nevertheless inform and influence the behaviour and feelings of the members. An organization is coherent to the extent that there is also a collective organization-in-the mind shared by all the members. In what follows, the terms 'organization' and 'institution' are used somewhat interchangeably. However, a distinction can be made between the relatively stable aims of something called an 'institution' with an emphasis on continuity and solidity, and the relatively more flexible and changeable connotations of an 'organization'.

Gordon Lawrence (1977) has shown that whilst an organization may have one publicly stated idea of its primary purpose or mission, there are often also hidden conceptions at work. Put simply, there is the level of 'what we say we do' but there are also the levels of 'what we really believe we are doing' and also 'what is actually going on'; the members of an organization may be quite unconscious of this third level (see Chapter 3). In Chapter 18, Anton Obholzer suggests that at this third level the health service is unconsciously seen as a 'keep-death-at-bay service'; whilst the stated task is the treatment of illness, there is also an unconscious task of providing each member of society with the fantasy that death can be prevented. Many of the extreme feelings about the provision of health services can be understood more accurately as having their roots in anxiety about death; death as one of the inevitable outcomes of hospital work is denied. In one case this actually took the form of a hospital in London being built without a mortuary. It turned out that the architect had forgotten to plan for this, and no one had noticed. The drive to preserve life as an organizational imperative then becomes dominant, often irrespective of the quality of life that the patient will have.

When the medical model of cure is transferred to the field of psychiatry, the result can be a denigration of ordinary care, often the only hope the extremely mentally ill have. What is idealized are the latest fashions in cure, which succeed one another with manic rapidity. The result is very self-destructive both to the organization's conscious primary task, which is disrupted by this more unconscious emphasis on the task of cure as above that of care; and also to the morale of the staff, whose abilities to care and to act realistically rather than omnipotently are consequently undermined. Since care is a slow process and does not provide the dramatic result desired, it is denigrated as being ineffective, whereas 'cure', which is exciting and offers a defence of omnipotent denial of the chronic nature of the problems, is idealized. This is also an expression of professionals' use of 'treatment' as a defence against the inevitable experiences of helplessness and failure when working with the severely mentally ill.

A second form of denial sometimes found in psychiatric institutions is a political analysis of mental illness which denies its reality and claims it is only a consequence of social inequalities. Rather than having to care for and treat ill patients, an often difficult and sometimes unsuccessful process, the staff are drawn into seemingly endless power struggles over irreconcilable ideologies. A frequent dynamic of staff meetings in such institutions is centred on who has the 'power' and who is 'powerless' within the team. This is a defensive shift away from the real powerlessness that the whole team shares in its relative inability to 'cure' the patient. For example, in a psychiatric unit to which I consulted there was, in the course of the staff meeting, a heated debate about a patient who had requested her social security cheque. The doctor's view was the patient was currently ill with a manic-depressive illness and would simply spend all her money on some useless article. The social worker disagreed; in his opinion it was a contravention of the patient's rights to withhold the cheque. There was a furious argument with the majority of the nurses and other staff supporting the social worker's view. The patient subsequently spent her entire cheque on alcohol and chocolates. I don't think it is possible to say that either the doctor's or the social worker's view was right or wrong, instead there was a painful choice – to restrict the patient's 'freedom' or to 'collude' with madness. One of the feelings that the whole team found difficulty in facing was their shared sense of helplessness in relation to this particular patient. A focus on action and quick decision was used as a defence against this painful feeling.

Current efforts to move the mentally ill out of mental institutions (which were, it is true, sometimes very cruel) often result in their being abandoned to live on the streets. In my view, this in part involves an unconscious attempt to solve the problem of currently incurable forms of mental illness through denial by allowing the sufferers to die from exposure in a hostile environment.

I am not arguing for a return to the large mental hospitals of the past, but it is necessary to acknowledge the complex emotions and often ambivalence surrounding the activities of cure and care. Unless this ambivalence is acknowledged and managed, and worked with rather than denied, there is a danger of a considerable amount of cruelty in any care system. Here, the underlying and unconscious institution-in-the-mind is one of a powerful cure for all mental illness, and hence each failure to do so is a threat to the identity and sense of effectiveness of each individual member of staff. This idea can become a persecuting presence and drive the staff to acts of rejection and cruelty towards those in care, which would not be the case in an institution with a less idealized picture of its primary purpose.

As a final example of the complexity caused by diverse pictures of the organization-in-the-mind of the members of an institution, I turn to the prison service. Here, there are at least three competing views of its primary purpose: to contain dangerous and violent criminals, to punish those who have broken the laws of society, and finally to provide rehabilitation. Different prison officers will emphasize one or other of these in their everyday work; likewise individual members of society will have different expectations of the prison service. This gives rise to conflict and confusion both inside and outside the prison service about the role of the prison officers. Are they to be controllers, punishers or rehabilitators? Any decision about action in relation to a particular person needs to take account of these conflicting and competing roles. Failure to acknowledge this will result in muddled and unreflective decision-making in the staff team, as well as stress for individual officers.

These different, and to some extent conflicting, pictures of the prison institution-in-the-mind and of the role of a prison officer need to be distinguished. Prison staff are of course aware of this, but nevertheless feelings about the priority of one purpose over another do often impede day-to-day decision-making, and also policy-making both at institutional and societal levels. Here, as for the other examples above, shifts from one to another view of the primary task can contribute to the felt need to 're-structure' organizations, as if re-structuring will solve the difficulties inherent in the work.

INSTITUTIONS AS CONTAINERS

I want now to return to how the individual member of an organization may be affected. Although often felt to restrict, constrain and limit the individual, institutions can also provide a sense of psychological and emotional containment. Much of the sense of constraint in organizations is produced because each individual member projects parts of the self that he or she does not want to be aware of into other, more distant parts of the organization. These not only provide a focus for blame for the frustrations and conflicts inherent in working in the organization, but also 'lock' individuals and groups into unconscious roles.

In all of us, there is the impulse to work and there is the impulse not to work (see Chapter 2). Where can this impulse against work be located? Very conveniently, in the department or office down the hall, or in the other building; *they* are the lazy ones, or the reason we are not doing well at the moment. This is one unconscious reason why we form and join organizations: to provide us, through splitting and projection, opportunities to locate difficult and hated aspects of ourselves in some 'other'. Internal personal conflicts can be projected on to the interpersonal or even inter-institutional stage. This also happens at the international and the inter-ethnic levels, but it is most immediately evident inside our own organizations. However, this process of splitting and projecting requires a reasonably coherent, clearly structured and relatively unchanging organization. If 'they' keep changing, how do we know who 'they' are? Stability is not a prominent feature of most organizations today; continual change and re-organization are in progress almost everywhere. Who is 'them' and who is 'us' has become less and less simple.

The current trend towards fundamentalist religions is perhaps related to this, reflecting a wish to find an exceedingly stable framework where one's actions are prescribed by a religious text, whether it is feeding, sexual behaviour or a ready-made and socially sanctioned enemy. In this way, anxieties about choices are reduced, which brings a reassuring feeling of stability, predictability and familiarity. The problem, of course, is that it requires an out-group to fight against, and in doing so one can provoke counter-attack from them. Whilst this may have been a tenable method of managing things in the past, particularly where the 'enemy' was always at some distance and less likely to be provoked, the increasing interdependency between nations and the rapidity of communications means that it is now a much less viable form of psychological defence. Additionally, national economies are now so interdependent that this psychological myth also becomes economically counter productive.

One of the things that a good 'old fashioned' institution provides is something that we can really love or something that we can really hate. And it will be there tomorrow, no matter how hard we love or hate it. Nowadays, one is hard-pressed to find such a thing. In the past, the institution could be blamed for being too conservative, too rigid, too bureaucratic or whatever, and very convenient that was too, since it provided the individual members with places to locate these conservative and rigid aspects of themselves. Nowadays, you will be lucky if the other department or the other unit is still there next year. Certainly the people may not be, and certainly the task is continuously changing. As a result, institutions are not so available for the working out and working through of the ambivalent feelings surrounding work that each individual has. It is often hard to be sure what one's organization will be doing in a year's time, and what its structure will be, and this causes anxiety.

The changeability of organizations means they do not provide such easy targets for projections. This results, I believe, in the widely shared experience of an increase in interpersonal tension and personal stress within sub-groups inside organizations, instead of the more familiar and simpler tension between workers and management. This is perhaps even more obviously true in public sector organizations, such as the health, education, social, civil and police services, than in commercial ones, though even the largest and most successful companies are changing out of all recognition. Whereas our public institutions used to provide a reliable and stable container for the nation, helping to manage issues concerning inequality, sickness and disorder, they no longer provide a dependable environment for its citizens.

One result is an obscuring of the psychological task of public sector organizations by an over-emphasis on cost-effectiveness. In the case of the police, the public are now called 'customers'. While no doubt an important change in the sense that the police are trying to view us in a different way and asking us to do likewise, this can bring with it a different problem. We need the police to be available, psychologically speaking, for the projection of certain of our attitudes towards authority. Indeed, accepting these projections, working them through, and handing them back to society at large is part of the task of the police. If the police see themselves only as providing a service, and do not realize that psychological containment of tensions within society is also a central function, there is likely to be an increase in disorder rather than a reduction. If the police are no longer available in this way to society, they will not provide the necessary sense of authoritative containment.

THE SHIFTING FOCUS OF ORGANIZATIONAL CONFLICT

With the increasing recognition of the plurality of our society, which contains many sub-groups each demanding representation and influence, the primary task of many established public sector institutions requires re-negotiation. Furthermore, existing authority structures are continuously being challenged, creating considerable additional stress and confusion for the members of these institutions as they attempt to adopt and modify their working practices to take account of these changes. The conventional model of hierarchical top-down organizations is being replaced by negotiations between sub-systems of organizations with fewer levels of hierarchy. Where previously authority was ultimately patriarchal and matriarchal in character, we are now seeing conflicts not so much with 'the authorities' but between sub-groups within society and within organizations. To extend the family analogy, many organizational conflicts today are more akin to sibling rivalry between brothers or sisters competing for resources and power.

One result of these changes is that difficulties that were previously managed by projection up and down hierarchical levels, or between established departments and units, may be forced down to the interpersonal level between members within an organization. For instance, there is a notable increase in 'bullying' in organizations (Adams and Crawford 1992), and other forms of scapegoating of certain individuals within organizations who are then subjected to intolerable pressures and are often driven out in one way or another. The individuals who experience this sort of treatment are very often at the boundary of two parts of the organization, like heads of departments, or at its boundary with the outside world, for example, receptionists or secretaries; these positions are particularly vulnerable to institutional pressures. No doubt the individuals involved unconsciously choose these jobs, which in some way suit their personality, but their personal characteristics are also fed by institutional demands and unconscious needs, as discussed in the next chapter. Sometimes there can be a happy match between the unconscious needs of the individual and those of the organization. An effective receptionist, for example, should perhaps have what Bion (1961) refers to as valency for fight-flight (see Chapter 2), which can be used appropriately in role in the work. But when the match is unhappy, or a job changes so that an earlier match is lost, the individual is at risk of breaking down. The question of responsibility for this is complex.

> Janice, the head of an adult education centre, came for a consultation regarding her role as manager. She was under considerable personal stress, and felt that her personal problems were causing her to fail as a manager. Although she was nominally the head of the organization, it turned out that she was really not allowed to use the authority appropriate to this role. She was undermined at every stage by the executive committee; when problems arose, they would not ask her in role as manager how she saw the problem or what the solution might be, but instead would set up innumerable working parties, task groups and so on. Ostensibly this was in order to 'involve' members of the committee in the work of the organization, but the effect was to undermine her authority. In addition, because the centre was in a multi-ethnic area and Janice was white, the committee expressed great worry about being accused of being racist, although there was no evidence of this. Indeed, Janice seemed to be a very competent manager. In the interests of 'enquiry' and 'democracy', the management structure had been undermined, leading to poor morale throughout the organization, and great difficulty making decisions.

It seemed to me that a good deal of the problem had to do with the internal power struggles and feelings of envy between members of the executive committee. These were dealt with by giving each member an 'equal' share of the responsibility. This disabled Janice from operating effectively, since she had a very limited range of choices. It usually fell to her to carry out only the more unpleasant tasks, such as sacking staff, the actual decisions having been taken by the committee, often without her involvement. The committee were also expressing their envy of her role by undermining her. The result was incoherence and confusion in both policy and organization. Through our work together, she was able to understand these dynamics better and to devise ways of tackling them through gaining the proper support of the committee chairperson, and ways of negotiating more appropriate delegation of authority to her by the management committee. Her symptoms of personal stress, ill health and anxiety disappeared. As a by-product, the chair of the executive committee also became better able to manage her own task, partly through advice from Janice.

Tolstoy wrote that he felt his freedom consisted in his *not* having made the laws. Perhaps what he had in mind was that, precisely because he was not responsible for creating the laws, he had a choice about what role he could take up in relation to them. The laws provided him with a framework, but not one that he was himself directly responsible for making, and thereby he had a certain degree of freedom even though constrained by these rules. But what are the rules in today's organizations? Often the only rule seems to be that nothing will stay the same.

Freedom comes not necessarily from changing the organization, which is too often seen as the only solution to almost any problem. It can also come from finding, making and taking a role in relation to the task and the structures available to support this (Grubb Institute 1991). Often, greater actual change is achieved in this way. Where adaptation is required, then changes should of course be made; but nowadays it seems that a new manager is not seen as having really taken up his or her post unless everything is re-organized. Re-organizing may have as much to do with the need to establish identity and mark out territory – the organizational equivalent of a dog urinating on a lamp-post – as it has to do with improving organizational functioning.

The result is that people, such as nurses in psychiatric institutions, who have been doing things in a certain way for many years are made to feel not only that this way is now outmoded, but that it was never of any use anyway. This inevitably corrodes morale. Change can be driven by a manic and contemptuous attempt to triumph over difficulty and conflicts. Feelings of compassion are cut off and the capacity for concern is projected on to others who are then seen as weak. The organization may well need someone who can represent or 'carry' weakness, such as a part-time worker or a member of a minority group. This individual is scapegoated and may even be driven out. Again, personal stress is caused by unconscious organizational conflicts, but because the conflict has been forced down to the individual and interpersonal levels, it becomes impossible to address.

CONCLUSION

In the face of constant change and chaos, we cannot either full-heartedly love or hate our institutions any longer, but we continue to be dependent on them, often far more so than we care to realize. Witness, for example, the sudden decline of individuals who retire or who are made redundant. An externally containing and coherent organization supports us, yet we also hate, envy and fear institutions for their apparent power over us, and they can easily become personifications of persecuting figures from our internal worlds (see Chapter 1). However, when organizations seem fragile and unpredictable, they become more like a rather inadequate foster parent than a second home. Then the inevitable feelings of hostility and envy towards parental objects, previously projected into management, have either to be denied or directed elsewhere, usually into intergroup and interpersonal conflicts, contributing to personal stress in organizations. This process can be compared to the somatization of internal conflict in the individual; when we are unable to deal with conflict at a mental level, it is pressed down into the body and finds expression in physical complaints.

Unless the management of organizations is sufficiently stable to be able to provide a clear definition of purpose and a reliable container for the inevitably ambivalent feelings of those they employ towards those in authority, then the organization will express its disorder through individual and interpersonal disorder in its members. This comes to replace a more appropriate and creative struggle with the task of the organization, a struggle supported by the structure of the organization. Interpersonal and intergroup conflicts can easily provide scapegoats, and the real problems remain unaddressed.

Alongside this continual clarification and working at the primary task of the organization, its purpose and aims, there is also the need for a parallel working at the roles best designed to carry out the primary task. As has been discussed, roles have both an overt and conscious aspect (for example, a job description) and a covert or unconscious element (for example, the flight-fight basic assumption underlying the 'unconscious job description' of a receptionist in an organization). Also, as in the prison service, there may be a series of conflicting and competing roles implicit in the organization. Unless management includes the management of opportunities for staff to understand these pressures, there will inevitably be an increase in stress at the personal level. Role consultation for managers is one way for staff to understand and work in these normal pressures and tensions of organizational life.

Chapter 14

The troublesome individual and the troubled institution

Anton Obholzer and Vega Zagier Roberts

We would all like to believe that the world is fundamentally a logical, well-managed place. Since the evidence against this is overwhelming it is inevitable that we seek a defence against finding it so frighteningly unsafe. A popular explanation, going back at least to the Old Testament story of Jonah, is that all would be well if only the evil ones, the trouble-makers, could be got rid of. Similarly, in institutions there is a great deal of blaming – between departments, between staff and management, or between the organization and the outside world, the politicians and policy-makers, whose wrong-thinking is the cause of all our troubles.

As discussed in the previous chapter, institutional difficulties are often attributed to the personalities of particular individuals, identified as 'troublesome'. Here, we will look at how such individuals are unconsciously 'selected' by the institution, and how both the individual and the institution can deal more effectively with what is troubling them.

TROUBLE-MAKING AS UNCONSCIOUS COMMUNICATION

Sheldon Road was a residential unit for sexually abused children. One summer the staff were feeling particularly pressed because several old-timers had recently left, and had been replaced by relatively inexperienced workers. They asked their manager, Nick, not to accept any more referrals until later in the year. Rather abruptly, there was an emergency admission of a nine-year-old boy, Terence. Nick explained to the staff that he had done all he could to avoid this, but that no other place could be found for the child, an explanation which the staff appeared to accept.

Around this time, Tony, the only part-time worker on the staff, told his colleagues he had seen the manager leave the room of one of the girls late at night, looking dishevelled. A week later he placed a formal complaint about this against Nick, adding that Nick had been coming to work intoxicated. There was an inquiry, the allegations were judged to be without foundation, and Tony was warned about the consequences of making unwarranted accusations.

His colleagues had been aware for some time that Tony was in difficulties. He had recently left his wife, had been coming late to work, calling in sick, and getting very behind in his paperwork. They were worried about him, and suggested he seek professional help, wondering among themselves if he might not be heading for a breakdown. After the inquiry, Tony became more and more agitated, and made even wilder accusations against Nick. Finally, he saw a psychiatrist, who prescribed medication and recommended extended sick-leave. The staff were angry and upset about what had happened, blaming Nick for reprimanding Tony too severely, but were also relieved that Tony, who had been unable to carry his share of the workload for some time, had left.

Just as things were beginning to settle down, the behaviour of the new boy, Terence, became a serious problem. He was quite abusive to staff, and on two occasions got involved in fights with other children which resulted in injuries. The staff protested to Nick that they were not trained to deal with this level of physical violence; they could no longer contain Terence, particularly now that they were more short-staffed than ever. Nick gave every support he could: extra meetings were held with the staff, additional supervision was given to Terence's key-worker, and money was provided for extra locum staff. However, Terence's behaviour continued to worsen. On the day that he threw another child off the swing so hard that she had to be admitted to hospital with concussion, Terence was removed from Sheldon Road and transferred to a psychiatric ward. The staff again breathed a sigh of relief; with their most difficult charge gone, they could 'get back to normal' at last.

Many groups and organizations have a 'difficult', 'disturbed' or 'impossible' member whose behaviour is regarded as getting in the way of the others' good work. There may be a widely shared belief that if only that person would leave, then everything would be fine. This view is very attractive, hard to resist and tempting to act upon. At Sheldon Road, the staff had lost confidence in their manager when he proved helpless to protect them from taking on another child at a time when they felt very vulnerable. Tony's breakdown expressed the vulnerability they all felt, and could be regarded as a strong unconscious message to management to attend to their difficulties. At the same time, staff needed to locate their vulnerability in one member: 'It is he, not we, who is breaking down.'

It is significant that the precipitating event, Tony's allegations against Nick, while unfounded, served to express the desire of all the staff to accuse management of misconduct, the real misconduct being Nick's untrustworthiness in failing to prevent Terence's admission. Tony was the best candidate to express this, given his family difficulties and argumentative style, and could be regarded as having unconsciously 'volunteered' for the task. And, as happens very often, no sooner had one troublesome person left, than another one appeared. When management failed to take sufficient heed of the first spokesperson, Terence was 'selected'. The child who was hurt by Terence represented all of the staff: 'Look what is happening to us as a result of your decision.' Terence's behaviour expressed not only his own rage, but also the rage all the staff were feeling at parental figures who abuse and fail those whom they should protect. At the same time, first Tony and then Terence were used to voice the group's unacknowledged anxieties about the quality of the service they were offering, anxieties which were split off, projected and finally got rid of by the removal of the 'whistle-blowers'.

SELECTING A TROUBLE-MAKER

This unconscious suction of individuals into performing a function on behalf of others as well as themselves happens in all institutions. For instance, the individual who gives the chairperson a hard time, holding up a whole committee by questioning and arguing every move, needs to be viewed not as a difficult or 'troublesome' individual – though he may well be that – but as an institutional mouthpiece, into whom all the staff have projected their disquiet. Sitting embarrassedly in the same room, signalling with their eyes that they wish to dissociate themselves from the trouble-maker, they are disowning that part of themselves which, by a process of projective identification (see Chapters 1 and 5), is located in the trouble-maker. So, too, with the chronically tardy staff member, or the reactionary, or the person given to violent emotional outbursts. Rather than seeing them merely as among the inevitable hazards of working with other people, we can more usefully regard their behaviour as a response to the unconscious needs of the institution.

> The manager of Links, an organization providing community support for the elderly, had been plagued for some time by the lack of competence in the staff he had inherited from his predecessor. This had been ascribed to their having been recruited in haste from among a very small number of not very suitable candidates. When vacancies arose and posts were advertised, there were dozens of applicants. Selection was a long, careful process, at the end of which the manager was asked by a colleague whether he was relieved to have the highly trained staff he had been longing for. 'Well, I'm not too sure about one of them,' he replied. 'She doesn't seem as clear as I would have liked about the kind of approach that is needed in community work.' Indeed, this new staff member – despite her excellent qualifications on paper – soon took up a group role of the confused, argumentative staff member who neither understood nor agreed with fundamental policies and procedures in the organization, the very same role her less well-trained predecessor had held before her.

At an unconscious level, the new staff member had responded to a job advertisement that read between the lines something like this: 'Wanted: volunteer required to voice the difficult, disowned, anti-task elements of the staff. Both internal and external candidates are welcome, but only candidates with suitably difficult personalities should apply.'

The difficult person unconsciously acts on behalf of all the staff, and the problem, rather than being attributed to personality, needs to be tackled on a group and institutional level. Instead of 'Isn't it terrible how X is behaving?', there needs to be a psychological and institutional move to 'We all have ambivalent feelings which we need to own, and those that relate to our work in the institution need to be taken up at work.' It is only from this position that some headway can be made in improving the functioning of the organization. At the same time, this helps the 'difficult' person shift out of the role into which he or she has been locked, as other members of staff become able to withdraw their projections. However, as it is often the most vulnerable or least competent group member who is selected to voice the dilemma, it can be all too easy for others to dissociate themselves from the spokesperson and to treat his or her behaviour as a personal problem. Identical processes occur between groups and departments, as described in Chapter 8, where the nursing and non-nursing staff were locked into antagonistic roles until the intergroup projections and underlying anxieties were addressed.

DEALING WITH INSTITUTIONAL DILEMMAS

Very often, the particular nature of the work of the institution determines the type and style of the problem that the 'difficult' person is asked to act out.

> Thorne House was regarded as a particularly progressive therapeutic community for the treatment of disturbed adolescents. It deliberately recruited staff with varied backgrounds and experience, on the basis that this would offer the adolescents a wide range of potential role models. Week after week, a great deal of time was taken up in staff meetings by rows between two of the men. Rodney was young, trendy and 'laid back'; he thought the staff should overlook what he considered minor infractions of the rules. Richard, who was older and had previously worked in a psychiatric hospital, on the other hand, insisted on taking these very seriously: 'If our boys don't know where they stand, things will get out of hand.' He was perceived by his colleagues as rigid and authoritarian, while Rodney was referred to as a 'bit of a flake'. The arguments between Richard and Rodney about how to manage the adolescents were endlessly repetitive. Other staff sat back and watched, some amused, some bored, and many irritated at the waste of time and the lack of change in either party.

The risk in situations like this is of seeing the process in terms of individual personality – or, in a therapeutic institution, in terms of individual psycho-pathology. This leads to a blind alley. It was not within anyone's authority at Thorne House to insist that staff have personal therapy. Even if it had been, it is likely that Rodney and Richard would have resisted, at least until they were freed from the institutional projections having to do with unresolved unconscious dilemmas, in both staff and residents, about issues of authority, managing themselves and being managed by others.

An intervention focusing on the institutional process could serve to draw all members of staff back into role, and enable them to resume work on the primary task of the institution. By contrast, an intervention focusing solely on the difficulties of the two individuals could produce only an unsuccessful therapy group, or a road show for a vicarious audience. Personal psychopathology is relevant in this kind of situation only inasmuch as it determines who will be used for what institutional purpose: their personal *valency* for particular unconscious roles (see Chapter 2). Seeing the individuals concerned as expressing a wider conflict directly related to anti-task processes puts the problem into a different framework.

Thus, what needed to be recognized at Thorne House was the two sides of an unexpressed institutional debate on permissiveness versus control, so that the fight could become the public, ongoing debate within the staff group as a whole that it needed to become. Indeed, this debate taps into the very essence of the adolescent process, with its unconscious struggle between authoritarian and anti-authoritarian parts of the maturing self. Similarly, at Sheldon Road, fundamental tensions in working with abused children – for example, between avoidance and intimacy, between the wish to expose and get rid of an abusing parent and the wish to keep the family intact, between trust and mistrust of both children and parents – were avoided. Instead, anxieties were projected into vulnerable group members with a valency for expressing one or other aspect of these dilemmas. As a result, Tony and Terence both behaved in such extreme ways that they were expelled from the institution, leaving the underlying issues unresolved and likely to re-erupt as a new crisis at any time.

From this group-as-a-whole perspective, troublesome individual behaviour must be perceived and treated as an important indication of a problem in the group. Thus, the adolescent who is caught smoking marijuana needs to be thought about and managed as representing drug-taking 'on behalf of the entire group'. Whether they are kept on or expelled, the meaning of the drug-taking behaviour needs to be taken up as an institutional issue, resulting from a complex network of projective processes. It is the most-vulnerable-to-drug-taking adolescent who is unconsciously selected to take the drugs; but merely to treat the individual is to guarantee that the problem will crop up again, in the same or in a different form. The difference in the two approaches to the situation is of major consequence. Treated as a group process, the underlying problem, as well as the individual's, is addressed. To treat it as one person's misbehaviour allows everyone else to continue disowning and projecting aspects of themselves into the targeted individual, and the process will continue unabated, to the cost of both the individual and the institution.

Institutional dilemmas, like personal ones, are anxiety-provoking, and regularly give rise to the kinds of defensive projective processes described above.

These processes can then lead to individual stress and scapegoating, as happened at Sheldon Road, and as described in the previous chapter. Alternatively, the unconscious roles particular individuals are pulled into may be fairly comfortable for them, as seemed to be the case for Richard and Rodney at Thorne House. In this case, the collusive lattice (Wells 1985), in which each member of a group accepts a tacitly agreed unconscious role, may continue indefinitely, to the serious detriment of the organization's primary task.

IMPLICATIONS FOR MANAGEMENT

At a minimum, each of us must manage ourselves in our various roles. This requires an ongoing awareness of the issues of tasks and boundaries (see Chapter 3) and of authority (see Chapter 4). It also requires awareness of institutional process, and our own particular susceptibility or *valency* to being drawn into certain unconscious roles on behalf of the institution-as-a-whole. Even just recognizing that one is being used – has been 'enroled', so to speak – to perform some unconscious task on behalf of others can be immensely liberating. This understanding also makes one somewhat less vulnerable to institutional processes.

Developing awareness of unconscious process

Transference and countertransference are both useful concepts in helping to make sense of how one is perceived and treated, and also how one feels oneself (see Chapter 1). Paying attention to our feelings, particularly when they are more intense than usual, may tell us when we are reacting to others in ways more determined by our past than by the present.

> Crystal, a charge nurse on Langham Ward, noticed she reacted defensively whenever Claire, the ward sister, criticized her in any way. She would feel anxious, hurt and angry, and would try to justify her actions in great detail, exasperating Claire with lengthy, argumentative explanations of trivial incidents. Crystal knew that Claire liked her and thought well of her work. Indeed, Claire had gone to great lengths to arrange for Crystal, who had worked on Langham Ward as a student, to work there after she qualified, and had supported her rapid promotion. So why, Crystal wondered, was she so defensive and anxious now?
>
> In therapy, Crystal talked a lot about her feelings towards Claire, how special their relationship felt when she was a student, and how upsetting she now found Claire's constant fault-finding. She came to recognize that she was re-experiencing aspects of her relationship with her mother, whose favourite child she had been. Her mother had been a pianist before her marriage, and had been very enthusiastic in supporting her daughter's musical talent. But when Crystal won a major prize and decided to study music professionally, her mother became very critical of her playing, and sceptical about her ability

to make a successful career in music. Her interest and affection seemed to shift to Crystal's younger sister, and Crystal became quite depressed for a period. Finally, she abandoned her musical studies to take up nursing.

Crystal realized that every time Claire criticized her, she reacted just as she had with her mother, feeling the old anxiety about losing her special place in the other's affections. The student role had been an easy, comfortable one, repeating the easy, close relationship she had had with her mother during her early years. However, the more proficient Crystal became, the more frightened she felt of arousing dangerous rivalrous feelings in Claire, and the harder she felt compelled to try to regain Claire's approval. Once she could stand back from her intense and irrational responses to Claire, seeing how they had more to do with the past than with the present, Crystal was able to respond more appropriately to criticisms, learning from some, accepting others as prompted by tension, and asserting herself calmly when the criticism seemed unfair.

Personal therapy can be very helpful in providing access to parts of one's past. As Santayana put it, 'Those who cannot remember the past are condemned to repeat it' (1905/1980). Even without therapy, however, one can work at developing a self-observing stance towards one's reactions, noticing when these seem more intense than the current situation warrants, or when one's emotional state is similar to ways one has felt in earlier significant relationships.

This can help, not only to understand and manage one's own behaviour, but also to understand and manage others more effectively. In the above example, even if Crystal had not changed, Claire might have recognized that Crystal was reacting to her in an overly emotional way, and suspected this was based on transference. She might then have handled Crystal's mistakes differently, or at least not have felt so hurt and exasperated by the changes in her protégée's attitude towards her.

Managing oneself in role

The risk in this way of thinking is that one may use knowledge about transference and countertransference defensively, disregarding every complaint made of one's behaviour on the basis that 'it is their problem, not mine', rather than examining one's own part in the difficulty. To guard against this requires constant monitoring of our own state of mind, and how this is determining our actions. Since we are in a much better position to change our own behaviour than that of others, insight into unconscious processes needs to be used primarily to manage ourselves.

The process is further complicated when projective identification comes into play, and one gets pulled into behaving like the person the other perceives us to be.

Students on a psychodynamically oriented management course complained that the tutors treated them in a harsh, uncaring and disrespectful way. The tutors found this frustrating and hurtful. They continually tried different ways of teaching, but it seemed to them that whatever they did was misinterpreted and distorted. It was puzzling, too, since in their other work contexts, even as tutors on other courses, they were regarded as supportive and helpful. They suspected that the students' perceptions of them were based on their experiences of their managers at work, whom they regularly described in identical terms. Since this was a psychodynamic course, the tutors shared this hypothesis with the students. This proved quite unhelpful; indeed, the students claimed this was but further evidence of the tutors' refusal to attend to their point of view, and of using their knowledge 'to put us down' rather than to be helpful.

Over time, the tutors began to notice they were discussing the students with each other in increasingly disparaging and judgemental tones. Whereas before they had been clear that something 'not me' was being projected, they now found themselves behaving and feeling in the very ways the students described. It was as if they had become the harsh, punitive managers they had been accused of being. At first the tutors rationalized this as a natural reaction to the students' seeming resistance to learning, but as they recognized the intensity of their punitive feelings towards the students, they realized they had become caught up in a process of projective identification. Only after they acknowledged their own part in this process, how they actually *were* being unreasonable in their expectations of the students, did the climate of the course begin to shift in a way that allowed the students to begin learning again.

Here, the tutors were helped to extricate themselves from the roles they had been sucked into by there being two of them, so that they had the benefit of each other's observations. Furthermore, their knowledge of unconscious group processes enabled them to recognize the 'numbing sense of reality', as Bion (1961) puts it, that accompanies taking up a particular role based on others' projections and becoming identified with the projected role. The impetus to reflect on what was happening came not only from the discomfort the role suction (Wells 1985) was causing them, but also from their holding in mind the primary task of the course, the students' learning, which was so obviously being impeded.

Re-framing the presenting problem

Besides personal insight, whether acquired through therapy or otherwise, it is immensely useful for managers and other professionals to have training in understanding group and institutional processes. Ideally, in addition to theoretical input, this should have an experiential component (see Chapter 4) so that they can become attuned to the interplay between their own personal valency for particular unconscious roles, and the institutional process. This can enable them to stand back from demands to sort out problems as expeditiously as possible, in order to identify the underlying issues requiring attention.

However, as institutions typically develop ways of functioning that serve defensive ends, and since institutions of a particular type tend to bring together staff with similar personal valencies, it may be necessary from time to time to get help from someone outside the organization. This may be an institutional consultant, or it may be someone whose role is sufficiently outside the immediate problem situation that they can help restore the capacity to think, which those inside have temporarily lost because they are caught up in the process.

Roxanne, an educational psychologist, kept getting referrals from Stepside Comprehensive School of boys who were accused of bullying. Each referral appeared sound enough in itself, and each invited her to give an opinion on what to do about the pupil involved. She noticed, however, what a steady stream of bullies were being referred, far more than from any other school in the area, and how few other kinds of referrals Stepside made. This raised for Roxanne the possibility that the referrals might be saying something not only about the children, but also about the school. It might be that the message was, 'We have this problem of bullying, therefore of authority and power, in this school.' The referred boys might be those with the greatest difficulty with these issues, who were being 'used' to deliver this message.

Instead of continuing to assess each boy who was referred, Roxanne suggested meeting with the staff at Stepside to look at what the bullying might be about. What emerged was a serious management problem. The headmaster was perceived as a bully, and in fact was using bullying tactics to manage the staff. Once this came out into the open, and with the help of individual role-consultation to the headmaster as well as consultation to the staff group as a whole to discuss these issues, there was a dramatic reduction in the number of referrals for bullying.

Because Roxanne was an 'outsider', she was able to notice a pattern in the referrals, rather than – as was happening within the school – seeing each incident of bullying as a crisis requiring immediate action. Her awareness of unconscious group processes helped her to intervene in a useful way, by tackling the problem as one belonging to the whole institution rather than to a number of individual trouble-makers. Work with the whole institution to take back its projections and deal with the institutional problem proved beneficial to the staff, the headmaster, the boys referred – and also to Roxanne herself, who was no longer inundated with more referrals than she could deal with.

CONCLUSION

The management role requires maintaining a position at the boundary between inside and outside (see Chapter 3). This applies both to managing oneself and to managing others. If one is too much inside – caught up in the internal group process or one's own inner world – one is likely to enact what is projected, rather than managing it, as was initially the case with Crystal. If one stands too far back, one is likely to lose touch with the important information the emotional experience can convey, or to use knowledge of group processes defensively. We have described some of the ways institutions use individual members to express fundamental institutional dilemmas, and how awareness of these processes can assist both individuals – who can begin to move out of the unconscious roles they have been locked into – and the institution as a whole to move forward.

Chapter 15

Facing an uncertain future

Francesca Cardona

Many organizations are facing an uncertain future. This understandably raises anxiety and stress levels, often to the point where people cannot think, and just feel helpless. It can also lead to greater difficulty in working together, due to distrust, lack of information, and greater competitiveness for dwindling resources. Yet in these situations, it is more crucial than ever to think, plan together, identify areas where one can still influence events, and hold in mind the needs of the clients. This chapter describes one example of this kind of situation, and how consultancy restored some capacity to face the future and make plans, despite uncertainty.

FIRST IMPRESSIONS

Green Lodge was a residential establishment for adolescents with severe behavioural problems, situated on the edge of a big city. Originally founded as an approved school, it consisted of two residential units, one accommodating up to nine and the other up to fourteen boys and girls, and a semi-independent flat for older adolescents preparing to leave the establishment.

The request for consultancy followed the re-opening of the smaller of the two units, and the consequent re-organization of children and staff. The management felt there was a need for reflection on the impact of these changes on the unit, which seemed to be experiencing considerable stress. When Peter, the deputy director, telephoned me, I was pregnant and quite reluctant to take on new work. However, he seemed so keen to establish some links with me immediately, even if the consultation itself would have to wait several months, that I agreed to meet with the senior managers of the community for an initial exploration of their request shortly before starting my maternity leave.

When I arrived, I was struck by the sheer size of the campus, the wide spread of the buildings, and the absence of a physical centre. What was immediately noticeable was how much the environment was lacking in character, and the absence of children around the place. The community seemed empty and isolated. The coldness of the place was in sharp contrast to the warmth of the first meeting with the management. They seemed eager to discuss their concerns, which centred on the stress experienced by the staff of the newly re-opened unit. I felt it was important to try to understand this within the context of the organization as a whole, and we agreed that I would send them a proposal for a consultancy project involving everyone working at Green Lodge, to be started in seven months' time. I wondered afterwards if my being pregnant had represented some hope of productivity, while at the same time allowing the management to express their ambivalence by postponing the beginning of a process which might bring some changes to the status quo.

THE PROPOSAL

It seemed to me central to help Green Lodge reach some shared understanding and 'diagnosis' of their difficulties. I therefore designed the intervention with three main objectives in mind:

- to provide an opportunity for staff from each section of the organization to make an initial assessment of their own situation in the light of the recent changes introduced by the management;
- to create a forum where people could exchange information, discuss ideas and share their feelings about what was going on across the organization; and
- to provide a framework for reflection and development that could become a permanent feature.

The format I proposed was therefore to have a number of meetings with each of the different sectors separately. These included the care units, the school, administration, management and senior staff. I would then have a final meeting with representatives from all levels and professional groups of Green Lodge. I saw this project as a *transitional structure* (Bridger 1990), where staff could begin a process of reviewing their organization, not only their aims and working practices, but also their relatedness to the rest of the organization and how the recent changes had affected this. It seemed important to create a sense of openness, and to make connections between ideas, feelings and doubts expressed in each of the various settings and different groupings. My expectation was that by the end of our work together we would have developed a shared picture of the community, and identified areas to be developed.

INITIAL MEETINGS

When I returned seven months later, I met first with the management group. I discovered there had been quite a high level of sick-leave in recent months. The union had been called in by the staff, who found it impossible to cope with the level of aggression and misbehaviour of some of the children, and were blaming the management for their difficult work situation. The managers felt there was a contradiction in the complaint of being 'abused' by the children, when the task of Green Lodge was precisely to work with and care for very abusive and difficult children. It seemed that over the months since my initial visit, the distress previously located mainly in one sector – the newly re-opened unit – was now pervading the whole organization.

Both the director, Clare, and Peter complained about the inflexibility of the local authority, and the difficulty of negotiating even the most trivial changes. For example, the recent setting up of a separate kitchen in one of the residential units, so that the residents could prepare their own breakfasts and snacks without using the central kitchen, was regarded as a major achievement. There had been endless negotiations and long delays before even this small breakthrough in the net of bureaucracy could happen.

I came out of the meeting overwhelmed by a sense of hopelessness and resignation in the face of a situation that management did not seem able to control. On the one hand, they were part of a complicated bureaucratic system which apparently governed and controlled even their smallest decisions; on the other hand, they had an angry and dissatisfied staff group who did not want to accept the very essence of their work. It seemed to me they lacked any clear sense of purpose to guide and support them in tackling all these issues and conflicts.

My hypothesis at this stage was that people were using the children, and their difficulties in dealing with them, to express a more profound sense of frustration with an organizational structure unable to give them adequate support or direction. It also seemed likely that the children's behaviour was fuelled by the unreliability and uncertainty of the authority structure. The emptiness I had been so struck by on my first visit, the absence of a physical centre, seemed to mirror the lack of a centred and coherent approach, and also the distance between a well-meaning, but ineffectual, management and an angry staff group.

On the same day, I met with the teachers. They were quite critical of the organization, especially of the lack of discipline, and seemed uneasy relating to the younger and less experienced residential care staff. Most of the teachers had been there for a long time and saw the school as the major strength of Green Lodge, providing an element of stability and structure for the children within an otherwise undisciplined and chaotic organization. They felt the school had been relatively unaffected by what was going on in the rest of the community. 'There have been dramatic changes outside in the last ten years, but I am still doing the same things,' summarized one of the teachers. It was as if the school were an island, safe and stable, but from which it was not easy to communicate with other parts of the system.

I met with each of the two teams from the residential units on my following visit. Team A gave a dramatic picture of their work. The children were very difficult and abusive towards them. Life in the community was almost unbearable for them and was not doing any good for the children. They seemed unable to think in a constructive way about possible changes and developments within the community. By contrast, Team B, which worked on the re-opened unit, arrived at the meeting very cheerful and friendly. They talked about the children with warmth, but complained that the children were all the same age and did not stay very long. They wished to create an 'ordinary', family-like environment where children could be given care, attention and stability, and they had many ideas about how to develop the community. However, they were finding it very difficult to do so. Certainly the units lacked homeliness, which seemed to reflect the transience, while also giving an impression of institutionalization.

INTERIM REFLECTIONS

Gradually I began to realize that the community was facing a very uncertain future. Although mentioned to me only in passing, the number of children was very low, with only eleven out of twenty-three places filled and not much hope of an increase in the number of referrals in the short term. Green Lodge had become a very expensive, under-utilized structure, which could not survive in an increasingly competitive marketplace. I also learned that the lease of the premises was due to expire in two years. What was surprising was that these serious threats to the survival of Green Lodge had not been identified earlier, nor indeed discussed as problems at all, merely as a background element to the conflicts and difficulties between management and staff, and between staff and children. As one of the managers put it, the potential closure was more a rumour than a stated fact. The re-opening of the second unit seemed to have been a decision unconsciously determined by the organization's need to deny the very real threats to their continued existence.

My slowness in putting the evidence together gave me a powerful indication of what was going on in the staff group. People were complaining about a number of things, such as the abusiveness of the children, the lack of discipline, and the bureaucracy of the local authority, but were not talking about what really bothered them: the possibly imminent closure. My own reluctance to face this reality, to say openly what was obvious, mirrored the attitude of the staff, who had avoided being explicit to me about the seriousness of the situation. It was natural for them to be protective about this issue with the external world, of which I was a representative; but they were also hiding it from themselves and each other.

I began to see that my main area of work had to be to try to help the organization to focus on its future. I knew that to accept the invitation to concentrate only on current working practices and internal changes would mean colluding in denying the main problems of the organization, but I felt painfully anxious about saying what needed to be said. I wondered if one of the reasons behind the secretiveness about the small number of children was related to a sense of guilt about their underwork, and anxiety about disapproval and punishment for not being cost-effective and not helping many children. Yet until people could name the danger which was threatening them, the anxieties it stirred up would continue to get displaced, probably into the children. Denial as an organizational defence disables people from thinking, and therefore from bringing about any change. Facing reality, on the other hand, can restore some confidence in their ability to influence their destiny, even if the external threat cannot be overcome.

My function therefore came to be one of disclosing this 'family secret', stating openly, on every possible occasion, facts that were known to all the community. As I did so, I became more and more in touch with the painful feelings of death and impotence which pervaded the whole community. I often thought back to how I had felt on my first visit to Green Lodge, how its sprawling emptiness, which so vividly reflected the essence of its problems, had produced in me a sense of depression that must already then have mirrored the experience of the staff working there.

FURTHER MEETINGS

As I continued talking with the different groups, I started to build up a picture of a very fragmented organization in which layers of different cultures and contrasting models co-existed with no real interchange between them. Most staff derived their professional identity from their past work experiences rather than from their present one. For instance, the teachers still related very strongly to their past experience of Green Lodge as an approved school, where there had been clearer boundaries, more discipline and less uncertainty about their task. Although they were against some of the brutality of the old system, most of them felt little commitment to the newer therapeutic community approach. The imprint of the past seemed too strong to be replaced with some frail, vague model of 'care'. In fact, all the children had committed some small crime, and the teachers saw the real task of the community as providing temporary containment for children who might otherwise end up in secure units and later in prison.

Other staff were also missing the clearer boundaries of the past. The administrative staff, and even some of the care staff, felt there was a lack of guidelines, that the system had changed too drastically towards the care end of the spectrum. For example, the night staff described a situation without any of the common sense usually found in a family, where children are expected to do some chores and to be accountable for what they do: at Green Lodge, even if children misbehaved badly, they were still allowed to go home for weekends. One residential worker, whose past work experience had been in secure units, told us of an incident where a child threw an object at him. He felt impotent because he 'could not lock him up', and was unable to think of other strategies to deal with the aggressive boy. Many staff complained about not feeling equipped to deal with such children.

Staff had very different perceptions of their tasks and roles. Some seemed to be working to a *warehousing model* (see Chapter 8), wanting the children to be dependent, well behaved, contained and containable, which they expressed in terms of a need for more discipline, a more formal organizational framework and secure units. They saw the children as fundamentally dangerous and in need of control. Others espoused a *horticultural model*, a more open approach to individuals' emotional needs, with an emphasis on 'normality' and close relationships with the children in family-like units. They saw the children as basically good but abused, in need of encouragement and nurture.

My impression, listening to all these views, was that their difficulties in dealing with the children had much to do with the absence of a clear and shared understanding of the primary task (see Chapter 3) of Green Lodge. Was the community a custodial institution with a caring component? Or was it aiming to create small family units where children could be looked after in a caring and protective environment? Outside the community, both options had been considered by the local authority, and either could have been implemented, but not both. As I came to understand it, the basis of this muddled sense of purpose was due to the combination of its origins as an approved school, with a bureaucratic model derived from the local authority. This had impaired the capacity to create a more independent, innovative and coherent approach; an underlying dependency culture affected the functioning of the whole organization.

The community provided for most of the basic needs of both children and staff: their meals were cooked for them, accommodation was supplied and the units were cleaned by domestic staff. In this situation, the staff seemed to have lost some of their ability to take initiative to plan and implement change, projecting it upwards in the hierarchy, and outside the organization itself. This attitude could also be seen as a projective identification with the residents, who did not feel in control of their own lives, and who both wanted and resisted containment of their impulses. As a result, staff and residents found it difficult to move towards a more active role, using their experience, resources and ideas in order to make an impact on their reality. Authority was experienced as distant and indistinct, but at the same time very powerful, and sometimes even abusive.

A further source of difficulty was the transitional nature of the organization. Children came here when other care agencies or foster parents had failed, and most did not stay very long. This sense of temporariness had always been a feature; even years before, there had been rumours about the expiry of the lease, though no one had sought to find out about it. All this contributed to the absence of a shared vision, philosophy of treatment, and planned approach to the development of the organization. As I saw it, no real transition from the former approved school system to a different organizational model and primary task had ever occurred.

THE FINAL MEETING

I decided to convey my message both in the design and contents of the final meeting, which I saw as a 'bridge' between the past and the future of the community, a structure which could become permanent and be used as a forum for planning strategy. The participants of the meeting included all members of the management group, together with representatives from each of the other sectors.

I started with a short introduction, focusing on what I saw as the main problem: the threat of closure, due both to the lack of referrals and to the imminent expiry of the lease. What did they want to do about it? After my introduction, people were divided into small groups; they had half an hour to discuss the possible closure of the community. This was designed to encourage them to think publicly on this subject, trying to identify together what were the main issues they would have to face if the community were to close. They wrote on big sheets of paper the main points to discuss later in the large group. They were then given another half hour to discuss the future development of the community, if it were to survive. Following this, the groups came together again for an hour to share their thoughts, concerns and feelings about the two possible futures of Green Lodge: closure or development.

I was struck by how anxious and lost people felt with regard to their future, and how difficult it was for them to plan and take responsibility in a working environment which they felt they could neither control nor influence. Many spoke of their worries about losing their salaries, accommodation and the protection of an organization. They were mainly preoccupied with themselves and seemed to have lost their sense of perspective. The anxieties aroused by the situation in the organization had interfered with their capacity to think about its future. The managers, on the other hand, were more in touch with the idea of potentially positive scenarios for the community in the outside world, but were unable to communicate these hopes effectively within the system.

After a communal lunch in the director's room, we re-convened to focus on planning. Staff were encouraged to look at how they could take an active role, whether in the death or in the survival and development of the community. Relocating the children well in advance and planning an ordered closure were discussed as possible positive actions if the local authority decided to close the organization. Identifying the strengths and potential of the community, trying actively to attract more referrals and using the large premises for a wider variety of programmes were considered as part of a strategy for development. The meeting was ended by Clare, the director, who summarized the issues which had emerged through the day and announced her intention to set up a working group to continue the work we had started. It seemed that being able to confront the threat of possible closure had made it possible to begin to think about ways of dealing with it.

CONCLUSION

In thinking about my intervention, and trying to analyse my countertransference (see Chapter 1) in relation to Green Lodge, I frequently wondered what I was trying to achieve in working with this organization. Looking back, I think I was trying to inject some energy and ideas into the system through the design of the project and through my own presence, fostering the hope of productivity which had been generated at the first meeting. My aim was to shake the dependency system and to help staff and management to exercise more authority over their situation and their future. Was it an impossible task?

Often, in the course of the consultation, I felt overwhelmed by the general sense of passivity, the lack of initiative, and the feelings of resignation expressed by the various groups. I felt negative, intolerant and angry with the situation, unable to see any positive outcome from my intervention, as Team A had felt about the residents. At other times, like Team B, I wanted to inspire hope, to convert the organization to a more active and purposeful way of functioning. Gradually, I came to terms with the fact that I was providing a facilitating and transitional environment in which to explore the issues related to their uncertain future, very like the actual task of the community. However, it was up to the management and staff to decide how and if they wanted to make an impact on their future – just as, in the end, it was up to the children what they would make of their lives.

The hypothesis that I developed during my work with the community was that only through an open and public acknowledgement of the real problem of the organization – the possibly imminent death of the community – could people begin to accept some ownership and shared responsibility for their future. Without this, very little could be done. I was not there to rescue the community, but to help them to find and use their authority (see Chapter 4), not only to influence the future of the community, but also to affect their job satisfaction and self-esteem, which in turn was linked with the future of the children. How could these young people learn to take some control of their lives if the staff felt they had no control over theirs? I felt that Clare's taking charge of the meeting at the end – which had not been planned – represented a major shift from the prevailing 'dependency culture', and indicated a new commitment to tackling the issues around the threatened closure. In other words, she had found her authority, which was the first step to the other members of the community finding theirs, and thus taking charge of their destiny and responsibility for their work.

It is often extremely difficult for organizations and staff to consider any change, even if change means moving on from a situation of great uncertainty and unhappiness to something possibly better, or at least more clearly defined. Change inevitably threatens established individual and organizational defences, and however unsatisfactory these defences are, they are hard to give up. The consultation described in this chapter was designed to contain anxiety enough so that the future, uncertain as it was, could be faced.

Chapter 16

Finding a voice

Differentiation, representation and empowerment in organizations under threat

James Mosse and Vega Zagier Roberts

In these days of widespread cuts, closures of services, and restructuring of jobs and management systems, many organizations feel under great threat. Often, they request consultation because they want help with managing the disturbances produced by these threats. In other cases, the request is for help with organizational development, such as training or team-building. Yet here, too, exploration often reveals a threat to survival, albeit an internal and unacknowledged one. We shall describe here a particular kind of institutional defence we have encountered repeatedly, which severely impairs an organization's capacity to deal with the threats to its survival.

AN EXTERNAL THREAT

As part of massive financial cuts across the whole education authority, the South Trenton Resource Development and Training Department, an advisory service for South Trenton schools, was about to be disbanded and its staff re-deployed. We were asked to design a training day on managing transitions, which might help the staff to cope with their anxiety about the impending changes. To help us prepare for the day, we were sent a mountain of documents; buried in these was a small diagram which showed that the proposed changes were going to affect some of the staff far more than others. No mention had been made of this when we were briefed, and when we asked about it, everyone professed surprise. Meanwhile, the trade unions had been mobilized, and on the day we met with the group, there was an excited fighting mood around. When one person said she did not feel like fighting, she was told they all had to stand united or 'they will pick us off one by one.'

The first exercise of the day consisted of a role-play. Participants were randomly divided into three groups: one to speak as the managers of the education authority, the second as the staff of the training department, and the third as the users of their service. Each group was to prepare a statement to present to the other two groups.[1]

---

1 We would like to thank Eric Miller for the design of this exercise.

The 'users' group was very energetic in criticizing the service, saying they had never been clear what it was for, and that they foresaw no great loss to the schools if it were to close. Up to this point, the value of the service had been treated as a given, the identified problem being that 'they' were not caring enough to appreciate its importance to the schools. That the 'users' could express their scepticism so quickly suggested to us that anxiety about whether the department really achieved anything had been near the surface, but not acknowledged openly and hence not dealt with.

> The second part of the day was designed to start a process which might enable them to respond more effectively to their situation. We asked each of the four teams within the department to prepare a proposal for alternative ways of cutting the budget of the service, and to select a representative to meet with the representatives of the other groups to prepare a joint proposal to the external management. This exercise became chaotic when whole groups insisted on attending the representatives' meeting, either because they had failed to select a representative, or could not trust one to work on their behalf at a distance. This made negotiation impossible, both internally and with the external management.

Here we see the interplay of internal and external threat. The depressive anxieties (see Chapter 1) about the usefulness of their work pre-dated the crisis for which consultation was requested, but had been suppressed: the cuts proposed by management may well have felt unconsciously like 'just deserts'. In defence, any blame of themselves or each other had been projected into the managers. This kind of shift, from feeling guilty to feeling persecuted, that is, from depressive to paranoid anxieties, makes it very hard for a group to think and act effectively. The pressure to band together and to blame problems on outside enemies was exacerbated in this case by the covert awareness that the proposed re-deployment would affect some of the staff far more than others. Indeed, all internal differences within the group had to be denied. This made it impossible to empower anyone to act or speak on behalf of the group.

The objectives of the consultation were two-fold. First, the group needed to begin to 'own' both the doubts about the usefulness of the service, and the need for financial cuts, rather than experiencing these as attacks from outside. In other words, they needed to face reality, to shift from basic assumption mentality (see Chapter 2), with its focus on personal and group survival, to work-group mentality, with its focus on the primary task, self-evaluation and survival of the organization in the outside world. The second objective was to help them begin to differentiate internally, so they could empower some of their members to negotiate on their behalf: to find a voice which stood some chance of being heard.

Away-days in times of crisis

Away-days have gained much currency over recent years, and have character-istics affecting both design and outcome. At the conscious, rational level, the 'away' component is about removing participants from their everyday work-setting with all its impingements, creating a space for reflection, debate and new thinking. However, the return to the office the next day is also a movement away – this time from the setting where the newness or learning took place, making it possible to leave these behind. If external consultants are invited to facilitate the away-day, the bounded timeframe of the contract means they will play no part in the process of implementing the away-experiences in the natural setting. They, and their ideas, can be used or forgotten according to the unconscious needs of the group.

In our experience, such away-days are often planned at times of crisis, whether acknowledged or not, with the participants having become locked into a particular stance in relation to the crisis. Some groups manifest a strong us-and-them polarization, 'fighting from the trenches' where they have dug themselves into defensive positions. Others are more obviously immobilized, waiting passively for the axe to fall. Typically, a fog of supposition and speculation replaces facts, which are neither sought nor recognized when available. The group seems to be paralysed by fear that any differences among the members will lead to fragmentation, or to sapping energy needed to resist the 'real' enemy.

In order to enable participants to shift out of the defensive positions into which they have become locked, we often use designs which contain an element of play, replacing the pressure to find immediate solutions with an invitation to 'play with ideas'. However, we set a real-enough task: not so real as to evoke the same degree of anxiety as the actual back-home situation, but resembling it closely enough for participants to have real investment in the outcome, rather than dismissing it as just a game of make-believe. The design usually includes an element of taking up roles different from members' everyday roles, both to encourage a shift in perspective, and because this requires participants to differentiate. Furthermore, the design includes some negotiation between sub-groups; this offers an immediate here-and-now experience of how difficult this is, and also a dramatic demonstration of how high the cost is of failure to differentiate roles sufficiently. Similarly, the disastrous consequences of excessive investment in consensus management and of inadequate delegation of power and authority are highlighted.

The whole design is intended to support a process of easy transition back and forth between playful as-if-ness and participants' back-home working reality. At its most successful, the day not only enables participants to begin to plan how to negotiate more effectively with other parts of their institutional system, and thus better to manage the external threat facing them, but also decreases the previous regressive pull towards dedifferentiation, so that individuals can engage in internal debate.

AN INTERNAL THREAT

The Felham Internal Counselling and Support Programme (FICS) had been established by the Felham Health Authority to provide practical and psychological help to any of its employees who might need it. Initially, FICS had a team of six counsellors, offering individual counselling sessions at the FICS offices. Later, a new project, staffed by a team of five experienced group-workers, was set up to offer consultation to staff groups complaining of stress. FICS staff had always had an annual away-day for development and training. For one such day, we were asked to design some team-building exercises to help integrate the two parts of the service.

We began by asking each team to list issues which were affecting the development of FICS. In the course of this discussion, we discovered that the group-work team had been set up as a one-year pilot project, which was unlikely to be renewed in six months' time because of cuts in funding. Having completed their lists, each group was asked to prepare a proposal for a training event to address the issues they had just identified. They were then to meet with the other group to negotiate a joint proposal for their next away-day.

The group-workers worked hard on the task we had set, while the counsellors spent most of their time arguing among themselves about whether or not they were obliged to do it. When the time came to negotiate a joint proposal, they brought only some vague, last-minute ideas. The other group's carefully worked-out plan caused them some surprise and embarrassment, but it was not debated. At the end of the day, it remained posted on the wall, unamended, neither accepted nor rejected.

By proposing nothing, the counsellors avoided the task of negotiation, since there was nothing to negotiate. The group-workers had put great effort into their proposal, despite the fact that they would almost certainly no longer be working at FICS by the time the next away-day took place. Having complied with the task, they had no further interest in their proposal, and left it on the wall to be removed by the cleaners. Thus, they too avoided negotiating with the other group. Our hypothesis was that acknowledging differences within the total staff group was intolerable, because of anxieties about guilt and envy in relation to the imminent demise of the pilot project. This significant event had not even been mentioned to us when we were briefed, and its importance continued to be played down throughout the day.

In our closing meeting with them, we pointed out that the issues they had listed in the morning were different for the two teams, and that some of them might be tackled more effectively if the teams worked separately, rather than hoping that their difficulties would disappear if only they all worked together. The evidence had been that when they tried to work together, the need to avoid acknowledging even the most obvious differences within the group made negotiation and debate impossible. Instead, anything either team came up with had to be forgotten or ignored to preserve unity.

THE PROBLEM OF DIFFERENTIATION

FICS was facing a similar situation to the one at South Trenton, namely, that it was about to be broken up and many of its staff re-deployed. Why had nothing been said to us about this when we were briefed? Why had we been invited to do team-building with a team that would soon no longer exist? Underlying the apparent differences in the situations of FICS and South Trenton, there were fundamental similarities. In both organizations, there was an external threat affecting some staff more than others. It was this, particularly, that could not be faced: both groups denied their internal differences. Both had invited consultants to help them bond even more closely together; in effect, to collude with their defensive dedifferentiation. In both cases, the denial of differences and avoidance of expressing clear views – which might conflict – prevented their negotiating ways to affect their future.

Threats to survival produce extreme anxiety. One of the commonest defences in groups under threat is to try to strengthen the emotional ties which bind them together. This includes denying any differences which could contribute to the dismemberment of the group. It is at such times that one is most likely to find groups under the sway of basic assumptions (see Chapter 2). South Trenton used basic assumption fight, demanding that its members join together to fight the enemy. FICS used basic assumption dependency, one team in a compliant way, the other by debating whether or not to depose us. Both looked to the consultants to provide basic assumption leadership by supporting their defensive dedifferentiation, and were very hostile to our efforts to identify differences, or even to name external realities that everyone was aware of but had 'agreed' not to voice. However, the inability to acknowledge and debate internal differences makes it impossible for a group to do effective work. The next example further illustrates this:

Argyll House, a day centre for physically handicapped people, had been working for some time with a small number of extremely disabled, dependent clients, who were grateful for any help offered them. The premises had just been enlarged in preparation for the centre taking on much larger numbers of less disabled people. The staff anticipated these new clients would be far more articulate, demanding and aggressive. Their request for consultation was framed in terms of shared anxieties, resistance to change, and difficulties in planning how to set up the new programme.

At our first meeting, it became clear that this was misleading. Some of them had found the old work tedious and were very much looking forward to more active and varied work in which they could mobilize skills for which there had been no scope before. Others were anxious both about the loss of the intimacy and predictability of the old work, and about lacking the skills that the new work was likely to require. In the meantime, they had been unable to plan their services for the new clients, who were due to arrive shortly. Ideas would be put forward tentatively, and quickly dropped when alternative suggestions were made.

One of the first interventions I[2] made was to ask the staff to work in pairs to

2 Where vignettes describe situations involving only one of the authors, we have used 'I' to avoid cumbersome alternatives.

prepare lists of the aims of Argyll House and of the activities they thought they should be engaged in with the different kinds of client, and then to order each list in terms of priority, or the amount of time to be allocated to each item. There was almost total match across the lists in terms of what was included, but extreme disagreement about the relative importance given to each item. I then asked them to produce a joint list before our next meeting: these would form the basis for their new operational policy.

Unacknowledged disagreements about priorities and about how the work was to be carried out had paralysed the team from making any decisions about the new shape of the service. Since neither consensus nor debate had been possible while the team felt so anxious about the potentially fragmenting impact of the new work, they had been unable to move forward. Now they were able to begin re-writing their operational policy. However, when they reviewed it six months later, they recognized it had troublesome shortcomings because many items had been worded too vaguely to be of guidance. This ambiguity resulted from their unconscious need to blur differences in the group. Instead of overt conflict, there was muffled antagonism in the team, which occasionally erupted in the form of personal criticism rather than as debate about the work itself. When the contentious items were more clearly defined, two team members who disagreed with what had been decided began to look for jobs elsewhere. This was initially perceived as evidence of catastrophic failure, as if it should have been possible to find a way to satisfy everyone. However, having clearly stated aims, values and policies made it possible to recruit staff who had a fairly accurate picture of what sort of organization they were joining, and of what would be expected of them. Instead of covertly sabotaging the work to move the group in one direction or another, team members were freed to manage their own work in relation to the agreed boundaries.

In groups under threat, the unconscious pressures on members to blur differences can be enormous – as if safety lies only in oneness. New members are required to make a tacit pledge to defend the group from any disagreement, lest this make the group vulnerable to attack from the dangerous world outside. Even the capacity to notice when something is wrong has to be suppressed. The price of belonging is the submerging of individuality, and is often experienced by members as oppressive, although they cannot locate the source of oppression.

One newly arrived member pointed out how much time was lost in meetings due to people arriving late. Two weeks later, although she still arrived on time herself, she no longer seemed to feel any of her earlier annoyance. When I remarked on this, she looked blank, leaving me feeling I was either mistaken or extremely tasteless to have recalled the matter. Similar blankness greeted my commenting that the black workers never participated in group discussions, or that team members received different pay and had different statutory obligations and different levels of expertise. When I persisted in these observations, the group would become quite hostile towards me; as soon as I stopped, the atmosphere would immediately become amicable again.

Gustafson (1976) has called this kind of group 'pseudo-mutual': roles are blurred, vague terms like 'helping' are popular, everyone must be 'equal' and 'equal' means identical. Everyone tends to behave in a friendly and helpful way. If anyone questions anything, this is likely to be met first by concern and offers of help, and later by hurt surprise at the unnecessary unpleasantness. The misguided member usually quickly relapses into anonymity. If they do not, they are likely to be extruded from the group with considerable, albeit masked, aggression. Under the guise of friendliness, there is actually a great deal of oppression.

Effective work requires differentiation: defining a clear task (see Chapter 3) and allocating work according to skills and resources. The pseudo-mutual group, instead of being held together by the work-based bond of a shared aim, is held together by a kind of 'glue' of identification. The group cannot bear separateness, ambivalence, imperfection and other sources of depressive anxieties, nor risk the emergence of envy and rivalry. Hence, there is no room for healthy competitiveness or genuine mutuality, only for friends and foes.

THE PROBLEM OF REPRESENTATION

A corollary of being unable to differentiate internally is that it becomes impossible to empower anyone to speak or act on behalf of the group, since this implies their taking up a differentiated role. Consequently, negotiation with outsiders is impossible. It may be possible to have a spokesperson, but only to transmit a prepared message, since the group cannot delegate sufficient authority to anyone to engage in dialogue with others on their behalf.

In addition, there may be an unconscious need to make sure that negotiations do not succeed, since the more dangerous the outside world is demonstrated to be, the more likely the group is to maintain the solidarity it regards as indispensable to survival.

> An employees' committee was set up at Shane Hospital as a vehicle for staff to express their grievances to the management. Each department was asked to send a representative of their own choosing. In almost all cases, since no one volunteered, the most junior member of the department became its representative. The committee was derided as window-dressing, and attending meetings was regarded as a pointless chore. Exploration of why the departments chose those least likely to be listened to suggested that there was unconscious investment across the hospital in proving that the invitation from management to set up the committee was not based on a genuine intention to take grievances seriously. When the nature of the relationship between the employees' committee and the management was debated and formalized, departments began to select more senior representatives, and the committee became an effective system for intergroup negotiation.

Effective representation requires thought about who is best able to take up the role of representative, based on explicit criteria such as being articulate, diplomatic or senior enough to be taken seriously by the other people involved in the negotiation. It also requires that groups delegate sufficient authority to representatives to enable them to pursue the task entrusted to them (see Chapter 20). This means being able to trust the representing member to act responsibly on their behalf. Yet often representatives are selected in an apparently random way. This serves to avoid both rivalrous feelings towards the person chosen and anxieties about differentiating a role, ensuring that nothing significant will happen. We have already seen how pseudo-mutuality at Argyll House led to suppressing individuation. Not surprisingly, this team also was unable to let anyone do work on its behalf.

> Whenever someone introduced a new idea at Argyll House, a working party would be set up to go away and think about it. Ostensibly, its task was to come up with a concrete proposal for implementing the idea, which the team as a whole would then consider. However, the invariable experience of these working parties was that when they brought back their plan, no one showed any interest in it, and the project died. With it, each time, died a bit more of the enthusiasm and commitment to the team of the individuals who had been involved in the working party, especially of the person who had first had the idea. Over time there were fewer and fewer individual initiatives, since their fate was so predictable.

In this case there was an apparent delegation of authority, a mandate to gather information and make a proposal. The outcome suggested that the creation of the working party was a way of getting rid of ideas, rather than of exploring them. A contributing factor was the membership of the working parties; since this was determined by who volunteered, only those committed to the idea were involved. Those opposing it could then kill it off by disregarding the proposal, without ever having to voice their objections explicitly, just as happened at the away-day with FICS (see pp. 149–51). The whole process served not to develop the service, but to remove contentious topics which might otherwise have threatened group unity.

CONCLUSION: WHO NEEDS TO SAY WHAT TO WHOM?

Threats to survival stir up primitive anxieties about annihilation and fragmentation. Very often, the response is to withdraw from reality, which seriously compromises the capacity for problem-solving. Some threats come from outside, as when an organization is at risk of closure or of being taken over. Others come from within, in the form of threats to self-esteem or to group cohesiveness. Very often there is an interplay of these different kinds of threats.

The question 'Who needs to say what to whom?' is often a useful prelude to planning how to manage the threat facing a group or organization. The 'what' part of the question needs to be tackled first, since in any danger situation it is essential first of all to recognize what the danger is. For example, at Green Lodge, described in the previous chapter, no one could say aloud what at some level everyone knew – that the institution could not be kept open with so many empty beds. Only when a state of mind had been reached where people could name the danger could they begin to think about what might be done about it.

The nature of the threat helps to answer the 'to whom' part of the question. For example, the external threat of funding cuts at South Trenton required negotiating with people outside the department, perhaps about alternative ways to achieve the necessary financial savings. Internal threats, such as the staff's anxieties about the usefulness of their work, on the other hand, needed to be discussed within the group. In many cases, this means bringing unspoken disagreements out into the open, and the group may well learn things which they would prefer not to know. Furthermore, internal debate involves relinquishing the fantasy that everyone will be pleased with the outcome. Usually there *are* winners and losers, but unless the group can bear this, everyone will lose (which may be psychologically less unbearable). It is crucial to have the right fight with the right people, so to speak; otherwise the fight will be displaced in ways that undermine the task.

Finally, having determined what needs to be said to whom, there remains the question of who will say it. Unless individuals can be empowered to speak, whether on their own behalf or on behalf of the group, the threatening conditions are unlikely to change. Denying the reality of internal differences is disempowering: neither internal nor external negotiation is then possible. In many cases, even the capacity to think will be lost. Yet now, more than ever, it is imperative to retain the capacity to think and act effectively under threat. If anxiety can be contained, then what needs to be talked about can be named, and some effectiveness recovered. Sometimes the threat itself can be overcome. Even when this does not happen, it is possible to regain some inner sense of having the power to affect one's own experience, rather than being a silenced victim.

Chapter 17

Asking for help
Staff support and sensitivity groups re-viewed

Wendy Bolton and Vega Zagier Roberts

When groups of staff in helping organizations themselves turn to others for help, one of their most common requests is to have a 'facilitator' for a meeting variously called a sensitivity group, a support group or a staff dynamics group. Usually they expect to meet weekly or fortnightly for one to two hours specifically to discuss stressful aspects of their experiences at work and how group members relate to one another.

The different names used for these sorts of meetings indicate some of the underlying assumptions. The request for a 'sensitivity group' tends to be based on the belief that 'getting things out in the open' is healthier than not doing so, and that becoming more aware of and sensitive to one another's feelings will prevent these feelings from getting in the way of the work. Since members are usually anxious about exposing personal feelings and perhaps being attacked, an external person is often brought in to keep things safe. The term 'support group' bespeaks the hope of getting more support from colleagues, as well as from the consultant, so as to cope better with painful aspects of the work. 'Staff dynamics group' shifts the emphasis more towards the usefulness of learning to be aware of unconscious processes within the group, and between staff and clients.

Elements of all three aims are found in varying proportions in most of the groups we have consulted to. The psychodynamically oriented consultant is likely to subscribe particularly to the third, and may well embark on the project in the expectation that the group will, over time, develop enough capacity for self-reflection and insight no longer to require the help of an outsider.

The value of regular discussions of painful experiences at work has been illustrated in many of the foregoing chapters. Yet, both the group members and the consultant sometimes find themselves involved in an activity which does not seem to be addressing the difficulties in a useful way. The meetings may stop, or hobble along for surprisingly long periods with considerable frustration, disappointment and lack of conviction on all sides. Our hypothesis is that such groups are often requested as an off-the-peg solution to an ill-defined problem for which they are not appropriate. Furthermore, even when a thorough assessment has been made and this kind of intervention chosen with thought, these groups are particularly vulnerable to counter-productive activity because of the unconscious aims and collusions of the participants, including the consultants. In this chapter, we will 're-view' how these groups are set up and used, and what can go wrong.

STAFF SUPPORT GROUPS IN PRACTICE: WHEN THINGS GO WRONG

The most usual criticism is that the discussions are repetitive and aimless, making no difference to how the work is done or how people feel. Occasionally they are seen as dreadful blood-letting sessions where feelings are expressed in an unthought-out and destructive way. Attendance becomes erratic, members drift in late or let themselves be called out early, and many of those present say little or nothing.

There are bound to be difficult times which the group has to experience before being able to understand and learn, and times when the consultant is almost overwhelmed by feeling useless, despairing or angry (see Chapter 7). The term 'facilitate' suggests the process will be made smooth, even easy, but more often it is slow and painful. Ideally, these experiences can eventually be used to increase understanding, with benefit both to members and to the quality of their work. None the less, it is important to take the criticisms seriously, for they may be an indication that the group has been drawn off-task, and that useful work is no longer being done.

Assessing the value or success of any activity is based on criteria, explicit or not, relating to the aims of the enterprise. Thus, members of these groups might judge their success according to how open members become with each other, how supported they feel, or how much awareness they develop of group processes and what use they are able to make of this. The situation is even more complex insofar as those involved – members, their managers (who have agreed to, if not arranged, the meetings) and consultants – may well have different aims, and therefore different bases for assessing their value.

We have found it useful to differentiate between *overt* aims, that is, those which are conscious and public; *covert* aims, those known to at least some of the participants, but not acknowledged; and *unconscious* aims, which remain to be discovered (Colman 1975). In the examples which follow, the titles of the subsections indicate the stated reasons for the requests for help, which include some of the most common problems for which support groups are set up. In our view, the groups were thrown off course largely because of participants' covert and unconscious aims, which we did not appreciate sufficiently at the outset. As a result, we fell into traps which we might otherwise have avoided. (Since the consultancy in each case was offered by one or other of us, not both, accounts are given using 'I' rather than 'we'.)

A problem of intractable conflict: Aspen Lodge

Aspen Lodge, a residential treatment centre for disturbed families, was in danger of being closed down because staff were not observing safety regulations. The families were cooking in their rooms instead of using the kitchen, and on one occasion had started a small fire. The management team, consisting of the director and three team leaders, felt helpless to bring about any change because of the entrenched animosity between themselves and the staff. A senior social services manager asked me to consult to a weekly staff support group in the hope that this would enable the unit to resolve their differences and stay open.

Little was achieved over the first few months. The atmosphere was of tiptoeing through a minefield, with people scarcely daring to speak for fear of being attacked, and the meetings punctuated by occasional violent explosions. Such discussion as did take place was about seemingly trivial matters, and the residents were rarely mentioned. I was increasingly unable to make any useful comment, overwhelmed by a sense of danger. To restore my own capacity to work in the sessions, as much as theirs, I began to consult separately to the three staff teams, focusing on their work with particular families. I also consulted to the management team on management issues.

In exploring the staff's experience of working with their clients, we came to see how the explosiveness and constant fear of violence in the families themselves was being projected into the staff group. The fire could be understood as an unconscious communication from the unit about danger, and the urgency either to get help or to be shut down.

Here the idea of meeting as a whole staff group mirrored the model at Aspen Lodge of working with whole families, as if managing to sit together in the same room for an hour and a half every week without destroying each other, or getting the families to survive under the same roof, would be enough to bring about the desired result. The overt aim was to have a forum where open discussion of feelings would lead to resolution of the antagonism in the group. As is typical in groups where anger and conflict are dominant symptoms, the unconscious aim was to get rid of all bad feelings. In situations like this, staff use the meetings as a kind of container which the consultant will take away with them when they leave, or as a black hole into which the badness can disappear altogether. It is then hardly surprising that members are reluctant to attend or participate, while adamantly insisting that the meetings continue.

When support groups meet for extended periods with little mention of clients, it is likely that the group has strayed into anti-task activity. At Aspen Lodge, it was by focusing on the work itself, and allowing the feelings the work stirred up to emerge from this starting-point, that we were able to make useful connections between the difficulties inherent in the nature of the work, the unit's working practices and the associated emotional experiences, in a way that led to thoughtfulness and informed change.

A team in crisis: the Westbridge Learning Difficulties Unit

The Westbridge Learning Difficulties Unit had been without a consultant psychiatrist for two years. As there was no team leader, individual team members took all their difficulties and disagreements with colleagues to the managers of their respective departments. They presented even minor problems with such a sense of impending crisis that their managers questioned the team's competence.

My first meeting was with the senior managers to discuss what they expected from my consulting to the team. Five out of seven of them were too busy to attend; the two who were there hoped that a staff support group would help the team manage its work and relationships better, without needing to involve the managers in the day-to-day running of the unit.

When I met with the team to explore what they hoped to get from the proposed support group, they complained bitterly of being neglected by management. As an example, they pointed out the naked light bulbs dangling from a broken ceiling, unrepaired since a storm several months earlier. They went on to say that what they needed was a forum where they could express their feelings about this situation. I suggested that while this might help them to feel better for a while, if the lights were still dangling in this makeshift way in six months' time, they would probably feel the meetings were useless. However, having waited a long time for the promised group, they were very keen for me to work with them, and I agreed.

Initially, the consultation seemed very useful. Most of the disagreements within the team had to do with lack of consensus about how their work should be done. They had postponed writing their philosophy and operational policy until the vacant post was filled, but now they undertook to draft this crucial document themselves, to be ratified at one of their monthly joint meetings with senior managers. As they engaged with this task, the team gradually allowed Daphne, their most experienced and articulate member, to take up a leadership role. Through the joint meetings with managers, to which I also consulted, they developed a number of creative new initiatives which considerably improved the unit's service to its large catchment area.

Then the Westbridge Health Authority was restructured, and during the transition the joint meetings with managers stopped. Soon after, Daphne died suddenly, after a brief illness. When no visit or even message of condolence came from management, the team sent a letter of outraged protest. The answer they eventually received was, 'That is what your support group is for.' After this, our meetings were again filled with repetitive complaints about the managers, the work with clients forgotten, and I became convinced that no useful work was being done any longer. I wrote a report confronting the management crisis, which I discussed at length with both the team and the managers, and arranged to draw the consultation to a close. Within a few weeks of my withdrawal from Westbridge, the long-vacant post was re-advertised and filled.

The overt aim of the managers who brought me in was to reconcile the differences within the team so that it could provide a more effective service. With hindsight, however, it is clear that their covert aim was that I should fill, or at least obscure, the management gap left by the vacant consultant psychiatrist's post, and reduce the additional demands it had put on them. The staff seemed to be seeking support to make this bearable, rather than looking for change, thereby colluding in denying the seriousness of the management vacuum. I had consciously seen my task as enabling the team to work as effectively as possible within the existing constraints through exploring their differences, agreeing a formal set of aims and policies, and allowing some natural leadership within the group to emerge. Since this seemed helpful for a considerable time, I did not realize the extent to which I was in fact also colluding in covering up the disastrous consequences of the situation.

This scenario, where a consultant is brought in to fill a management gap, is surprisingly common. Almost more often than not, we come on the scene to find either that the manager's post is already vacant, or that his or her departure is announced as soon as the consultation gets under way. In other cases, management is so weak as to create a functional vacuum, as was the case of Cannon Fields, described later in this chapter. The covert invitation to take up an unofficial management role can be seductive, and the consultant must be constantly alert to the risk of getting drawn into this anti-task activity.

Coping with change: Pine House

The staff at Pine House, a social services day centre for people with disabling emotional and psychiatric problems, were having to manage the arrival of several new staff, including a new team leader, and also imposed changes to their working practices which were seen as controversial, at a time when the team were already feeling unsettled and unsafe. The previous team leader had left after a prolonged acrimonious and unhappy time, leaving the staff feeling angry and guilty, wondering whether they had made him ill.

I was struck by the discrepancy between the bitter and vitriolic arguments in our staff support group meetings, and the reported ease with which the same staff had worked together on a planning day to reorganize the service. However, the decisions made on that day, and also at their weekly policy meetings, were repeatedly forgotten or sabotaged. I therefore suggested that I attend the policy meetings, as this might help us understand what was happening.

At each policy meeting, a different member of staff would open the discussion of the issue to be considered that day by presenting some prepared ideas. The same person would chair the meeting, which was difficult to do while also presenting his or her own ideas. Recording was left to a new clerk who was not familiar with minute-taking and knew nothing of the issues involved. Afterwards, there was little agreement or even memory of what had been discussed or decided, and the minutes were of little assistance. The staff were surprised when I pointed out these obvious incompetencies. The staff were certainly not incompetent, and it was a powerful lesson for them to recognize how the meetings were unconsciously being disorganized, so that they could avoid facing their disagreements about the new policies. When the staff began to be able to express their disagreements openly, and to record, remember and live with the consequences of the decisions they made, the outbursts of personal hostility in the staff support group became far less frequent.

Although the expressed aim for the staff support group was to cope with all the recent changes and to find a way of working with the new team leader, the covert aim was to avoid conflicts over policy in the decision-making meetings, 'holding over' differences for the staff support meeting, where they erupted in a personalized way, divorced from the work, which resolved nothing. The unconscious aim seemed to be to make sure there would be no winners and losers, so that no one would be hurt or angry at being forced to submit to unwelcome decisions; above all, it was that they should not damage the new team leader, as they feared they had the previous one, or be harmed by him.

When staff are working with the kinds of difficulties, frustrations and pain inevitably involved in caring for very distressed clients, it can be very useful and appropriate to have a group meeting separate from policy and business, to explore some of the intense feelings stirred up by their work. However, when the two types of meeting are used to compartmentalize feelings and action, the support group is almost certainly being misused.

Feeling isolated: the Beeches

At the Beeches, a day centre for the elderly, the staff requested a support group at a time when the centre was coming under criticism from outside agencies and the workers were feeling confused and dissatisfied at having lost their earlier sense of direction. This group was a particularly friendly one, yet the members' main complaint was that individuals felt terribly isolated in their work. Why was this so?

Their work was difficult and offered few rewards, as the ageing clients tended to get worse rather than better, and were often demanding and angry at their increasing helplessness. The staff, too, had to struggle with their own feelings of helplessness in the face of being able to do so little. In meetings of the staff support group, the feelings expressed were mainly of sadness, loneliness and loss. For example, the team returned again and again to their feelings about a long-serving member of staff leaving. This was certainly an important event, but it began to be used defensively, as a vehicle for sharing acceptable feelings while avoiding other, less acceptable ones, such as their anger at clients, their resentment of giving so much while getting so little, and their doubts about being of any use.

As at Aspen Lodge, clients were hardly ever mentioned. In fact, some members openly referred to the support group as 'something for us, not for them'. Despite their apparent cosiness and supportiveness, the group meetings did not help them to feel less isolated, or to work more effectively.

The overt aim of the support group was to provide a space where staff could discuss difficult feelings in order to regain a sense of direction and purpose. The covert aim, however, seemed to be to have time out away from the clients, the demands of the work and the feelings it stirred up. The criticisms from outside made it all the more difficult to risk having any disagreements within the group, and these were carefully avoided. Individuals were unable to acknowledge to themselves, let alone to share, the most difficult feelings, those causing guilt and shame, lest these further erode their already shaky self-esteem. This increased their isolation and further disabled the group from actually thinking about their work. As Hanna Segal puts it, 'In the same way in which the fear of an external authority can make us afraid to speak, the fear of an internal authority can make us afraid to think' (1977: 219).

Using a support group as time out is anti-task, and the consultant needs to help the group to think and talk about their feelings in a task-related way. At the Beeches, helping the team see how their clients' anxieties, resentments and confusion 'got into' the staff enabled them to begin to declare not only their feelings, but also their criticisms and disagreements about how the work was done. Only then could they work together on changing and developing the service. Insofar as their isolation stemmed in part from the nature of the work itself with painfully lonely elderly people, it persisted; but the sense of being part of an effective team offering a good service, and being able to share the painful anxieties the work stirred up in them, helped to alleviate their isolation considerably.

Fostering effective teamwork: Cannon Fields

Unlike all the other examples discussed above, Cannon Fields, the community mental health centre already described in Chapter 3, ostensibly had no problem. Indeed, the multidisciplinary team contacted me about facilitating a sensitivity group even before the centre opened.

At our first meeting, we could scarcely hear one another because two workmen were hammering at the window and peering in at us from outside. Yet everyone carried on with the discussion as if this were not happening. When I wondered aloud if I were the only person bothered by this, they explained that they had no authority over the workmen, who had been sent by 'central admin' at the hospital. It was only after I expressed scepticism about their being as helpless as they felt that one member went out and arranged for the men to work on another part of the building.

They explained that the new centre had an open-door policy, where any distressed person in the community could come for help, without any of the repressive rules or rigid procedures which had so impaired the quality of life for both patients and staff at the hospital. The discussion then moved on to the problems caused by one client who was arriving at the centre every morning drunk and abusive. That morning, he had so intimidated another client that staff had had to ask him to leave. Should they have a policy about drugs and drink? But what about the policy of being open to everyone? And who had the authority to decide? Finally, the risk to other clients overcame their reluctance to have rules, and they agreed that no one would be allowed in the centre while under the influence of drugs or alcohol. For nearly a year, this was the only formal policy, and the only decision which could be remembered and told to newcomers.

The team also found it almost impossible to impose decisions on each other, even those made by consensus. At our third meeting, a newly arrived staff member lit a cigarette. No one said anything to her about the decision made with great difficulty, and taking up over half the session the previous week, that there would be no smoking in these meetings. She became visibly more and more uncomfortable and eventually put out her cigarette, as if absent-mindedly, having smoked only half of it. When I commented on what seemed to be going on, it turned out that only one person besides me remembered that a decision had been made.

While there were no formal rules, the group seemed to be more and more inhibited by a growing body of unstated ones. One of the most powerful was that everyone was equal, with equal responsibility and expertise. It was very hard for anyone to raise any issue troubling them unless it was equally troublesome for everyone else. As a result, many sessions were taken up with repetitive, aimless grumbling about senior management, the one aspect of their experience which they could all agree was a problem.

The stated aim of the sensitivity group was 'to promote effective teamwork so as to provide a good service'. However, the unconscious aim of the group became to obliterate all differences between individuals, disciplines or hierarchical levels. One immediate consequence, evident already in the first meeting, was that no one could take action on behalf of the group (see also 'The Problem of Differentiation' in Chapter 16), nor could decisions be made except by consensus, lest some members appear to be more powerful than others; even then, the decisions could not be binding on anyone who had not participated in making them. Intended to be liberating, this state of affairs proved to be very constraining. Since one cannot oppose rules which do not exist, team members felt oppressed without knowing why.

Although there was a designated team leader, she could not be allowed – or allow herself – to have any power or to exercise any authority. Functionally, therefore, there was a management gap. With hindsight, I think I fell into the trap of trying to fill this gap, discussing the group's policies rather than working more with them on their profound ambivalence about the 'parent' hospital (on which they depended for referrals, resources and support) and its authority. This ambivalence mirrored the unconscious conflicts of their most difficult chronic clients, who continually demanded help while refusing to engage with the team in any treatment programme, and kept getting re-admitted to the hospital they both hated and needed. Certainly, the team's almost total avoidance of discussing their actual work with clients should have alerted me much earlier that we had strayed from our task.

CONCLUSION

All groups under stress tend to resist change, to seek 'magical' solutions and to collude in flight from their task (see Chapter 2). Support groups can help contain the anxieties stirred up by the work, restoring the capacity to face reality, without which effective work is impossible. Through exploring their work experiences, members can come to recognize counter-productive defences, question practices previously taken for granted and feel less isolated. However, when the difficulties derive from defences and other problems in the wider institution, a staff support group alone will not be sufficient, and can even be harmful.

In this chapter, we have illustrated some of the unconscious processes which can undermine the potential value of support groups. In some cases, the problem probably required a different kind of intervention altogether. For example, support groups are unlikely to be appropriate for dealing with crises, or with the consequences of absent or inadequate management. Another common mistake is to use such groups to do away with differences among participants, whether differences of role, authority or point of view. Probably the most frequent mistake of all is to separate work meetings from support meetings, that is, to divorce feelings from practice. As Hornby puts it, what is needed is 'task-orientated staff groups, with a structure and a group ethos encouraging open discussion of those personal feelings which are connected with work-role' (1983: 49). She added that support is needed to enable staff to face rather than evade difficult issues.

When a consultant is asked to facilitate a staff support or sensitivity group, it is essential to start by making as thorough and open-minded an assessment of the group's problems as possible before agreeing the method of intervention. Part of this process will be to clarify as much as is possible the overt, covert and unconscious aims of all concerned: the staff, the managers and, not least, the consultant's own. If a support group is chosen as the appropriate working method, these aims will need to be reviewed continually. Such reviews can alert all participants to difficulties and counter-productive collusions, and at times may suggest ways of changing the intervention, as in some of the examples above.

The projective identification processes inevitable in the course of constant intimate contact with very distressed and disabled people, and the risks these pose to care staff, have been described in a number of earlier chapters, especially Chapters 5 and 7. One can regard these as the 'toxins' to which workers are exposed, and support groups (when they are successful) as providing a kind of 'dialysis' to remove these toxins so that workers can continue to function without undue harm. The need for such a dialytic process is likely to persist as long as the exposure continues, so that the expectation that successful consultants will eventually make themselves redundant may be unrealistic. (The constant turn-over of staff makes it all the more unrealistic.)

However, consultants need to be alert to the support group being used to process toxins which are not inevitable, but rather products of inadequate management, organizational structures and support systems. Otherwise, they may find that they are working at making bearable what should not be borne. They must also be aware that these same toxins can 'get into' them too; they will need support systems of their own to contain their anxieties and help them make sense of their experiences.

The pull towards getting caught up in unconscious group and institutional processes, using groups to meet one's own needs rather than to further the task for which the group exists, is universal. Only if the consultants can disentangle themselves sufficiently from these processes to think, be aware of their failings without too much guilt or need to blame others, and maintain a reflective attitude towards their own feelings and behaviour as well as towards the experience of the group members, can the group develop a similarly thoughtful, non-judgemental, self-scanning stance.

Part IV

Towards healthier organizations

INTRODUCTION

Institutions are not always in crisis. There are, however, always organizational tensions which need to be managed. While many chapters in foregoing parts of this book have included a section suggesting how the particular issues raised might be managed, this part examines some features of organizations that need to be understood and managed at all times.

Chapter 18 puts forward the idea that our large public sector institutions serve as containers for fundamental human anxieties. The health service with its function of protecting society from anxieties about death, and the education service with its function of protecting us from anxieties about the future of our children, are the chief examples discussed, but the conclusions drawn about the basis for sound management apply equally to all public sector organizations, and indeed to many others.

Chapter 19 examines the tension endemic in the human services between care and control, particularly as it manifests itself in the supervisory relationship between managers and those they manage. This relationship can and should be a central source of support for staff. It also provides a mechanism for monitoring the quality of their work. Successfully combining these two aims within the supervisory relationship is a key to healthy organizational functioning.

Chapter 20 considers the difficulties which arise between agencies, or between departments and groups within one organization, or among members and sub-groups within a single team. Intergroup projections occur at all these levels, contributing to many of the conflicts which plague organizational life. Using an open systems model, together with ideas about the exercise of authority as discussed in Chapter 4, this chapter outlines some ways of managing intergroup relations more effectively.

The last chapter focuses on evaluation. While there are currently new pressures on most helping organizations to assess performance and outputs more systematically, some form of evaluation is essential for any work group. Without seeking to know about its effectiveness, a group inevitably loses its capacity to adapt and develop. While describing one particular approach to evaluation, this chapter underlines more generally the importance of fostering a spirit of enquiry within organizations, so that evaluation can become a tool for learning from experience.

Chapter 18

Managing social anxieties in public sector organizations

Anton Obholzer

It is hard to avoid the conclusion that what is wrong in our public services (or public sector organizations) today has to do with management. Article upon article, broadcast after broadcast, inform us that financial criteria have not been adhered to, that workers have too much power, that working practices are outmoded, that we are not as successful a country as we might be. The implication is that we are too soft: we need firmer management based on sound economic principles, and we need less consultation and more action.

No doubt some of this is true. However, as a psychoanalyst and consultant to institutions, I am struck by the parallel between human psychic processes and institutional processes. In individuals, it is a recipe for disaster to ignore the underlying difficulties, be they personal, marital or familial. Managing them by denying and repressing them invariably leads to further difficulties and disturbances. To avoid the underlying difficulties in institutions and to just 'manage' them away has similar consequences. Awareness of underlying anxieties and fantasies enables us to manage ourselves and our systems in such a way as to make improved use of resources, both psychological and physical. It follows that neglecting to do so results in disproportionately heavy wear and tear of both human and physical resources.

Management, structure and organization are not unimportant – in fact, they are vital. Nor is the emphasis on money inappropriate: financial constraint is a reality. In my experience of consulting to institutions, however, I regularly find that no attention whatsoever is paid to social, group and psychological phenomena. Consequently, by neglect, the factors that should be an integral part of good management become the very factors that undermine the venture.

As an example, it is common knowledge that any group numbering more than about twelve individuals is ineffective as a work group, incapable of useful debate and effective decision-making. Yet a great many committees are made up of many more than twelve people. One assumption might be that their purpose is not one of decision-making, but ornamental. An alternative postulate would be ignorance – that it is not known that groups of a certain size have certain dynamics, and are only capable of certain tasks and not of others. If so, it confirms the lack of group-dynamics knowledge in management. A third possibility is that such groups are unconsciously set up this way to ensure work does not get done, an anti-task phenomenon (see Chapter 3).

Besides group size, other factors necessary for groups to be effective include clarity of task, time boundaries and authority structures. And yet it is quite common to receive the agenda for a so-called work group too late for it to be of any use in preparing oneself, for meetings not to start on time, and not to be clearly chaired. Not only are groups frequently too large for work, but it is also common for their membership to be so inconsistent as to make work impossible: if one representative cannot come, another is sent instead. Eventually it reaches a stage where individuals come and go and no one is sure who people are or what they represent. This makes for great difficulty in maintaining an ongoing strand of work. It is possible to see why representative groups have got a bad name: the conclusion can easily be that representation and consultation do not work. The problem, however, is with numbers and structure, not with the process of consultation; it is to do with how the group is constituted and how it is managed.

The recent emphasis on 'tighter' management and control is a response to these kind of processes. The previous systems of public sector management are understandably written off, but the new concept seems to be based on a lack of understanding of what went wrong in the earlier scheme. So we have repeated re-organization, each equally uninformed and unsuccessful. These changes, directed at improving organizational effectiveness, come from what Reed and Armstrong call 'purposive systems thinking', which focuses on input–transformation–output processes. However, effective management also requires 'containing systems thinking'. This focuses on how people's needs, beliefs and feelings give rise to patterns of relations, 'rules' and customs which often continue unaffected by structural changes (Grubb Institute 1991).

INSTITUTIONS AS CONTAINERS OF SOCIAL ANXIETIES

Here, I use the term 'institution' to refer to large social systems such as the health, education and social services. Each of these, besides providing for specific needs – health care, schooling and so forth – through its primary task, also deals constantly with fundamental human anxieties about life and death, or, in more psychoanalytic terms, about annihilation. As discussed in earlier chapters (see Chapters 1, 5 and 7), the individual who is prey to these primitive anxieties seeks relief by projecting these anxieties into another, the earliest experience of this being the mother–baby relationship. If all goes well, the mother processes or 'metabolizes' the baby's anxieties in such a way that the feelings become bearable; we then say the anxieties have been 'contained' (Bion 1967). It is this process of containment that eventually makes possible the maturational shift from the paranoid-schizoid position, which involves fragmentation and denial of reality, to the depressive position, where integration, thought and appropriate responses to reality are possible. In an analogous way, the institutions referred to above serve to contain these anxieties for society as a whole.

Health care systems

In the unconscious, there is no such concept as 'health'. There is, however, a concept of 'death', and, in our constant attempt to keep this anxiety repressed, we use various unconscious defensive mechanisms, including the creation of social systems to serve the defensive function. Indeed, our health service might more accurately be called a 'keep-death-at-bay' service.

All societies fear death, and a multitude of systems exist in every society in order to cope with this anxiety. Some may attempt to cope with death by viewing it as a form of continuation of life. In many religions, belief functions as a socially sanctioned form of denial: you will not die and be nothing; dying is merely a transition, a step on the path of life. However, it is not only religion that is called upon to protect us from our most primitive fears and fantasies of death: doctors have always belonged to a similar defence system.

Originally, priests and doctors were often one and the same profession. While in Ancient Greece the division took place early on, in other societies the two strands are still located in one person, for example, the African witch doctor. In many rural or isolated communities, priests are still assumed to have healing powers, and doctors need to have some 'magic' as part of their practice. Although the practice of magical rituals has waned in our so-called civilized world, we would seriously mislead ourselves if we believed they are no longer of significance. I believe that many of the organizational difficulties that occur in hospital settings arise from a neglect of the unconscious psychological impact of death or near-death on patients, their relatives and staff. Hospitals are as much an embodiment of a social system that exists to defend society and its citizens against anxieties about death as are churches; from a psychic point of view, doctors occupy a similar niche to priests.

In some countries, there is a national health service which is used as a receptacle for the nation's projections of death, and as a collective unconscious system to shield us from the anxieties arising from an awareness of illness and mortality. To lose sight of the 'anxiety-containing' function of the service means an increase in turmoil, and neither its conscious nor its unconscious functions are served adequately. Consider, for example, the outrage in developed countries when advanced medical technologies cannot be made available to all; or the unfounded hopes placed in experimental treatments; or the tendency to feel duped when interventions fail. In all these situations, both individuals and society at large are quick to blame, as if good-enough medical care should prevent illness and death. Patients and doctors collude in this to protect the former from facing their fear of death and the latter from facing their fallibility.

Education systems

All societies have an 'education service', in the broadest sense, in order to teach their members to use the tools they need to survive. From an unconscious point of view, the education service is intended to shield us from the risk of going under. It is also, therefore, an institution that is supposed to cope with – whether by encouragement or denial – competition and rivalry. The debate about which nation has the best education system could be seen as a debate about who will survive and who will end up against the wall.

Institutions often serve as containers for the unwanted or difficult-to-cope-with aspects of ourselves. One source of anxiety in our society is our sense of responsibility for bringing up our children, and for their learning the skills needed to survive in society. Put this way, it is a fearsome responsibility, which, if put on to 'them' – teachers, schools and government education departments – lets us off the hook. For the office-bearers of the system, this is a double-edged sword. On the one hand, they welcome the power that comes with the job; on the other, the responsibility is terrifying, particularly as the expectations cannot be met.

At an unconscious level, what is hoped for from the education system is unreality: that all our children will be well-equipped – ideally, equally equipped – to meet all of life's challenges. For example, when I worked as a consultant to Goodman School, a school for severely physically handicapped children, the head welcomed me with, 'In this school we treat all children as normal.' While at one level an admirable statement, this encapsulated the denial of the extent of the children's problems and hampered all attempts to deal with them. The teachers had been trained in and operated in the belief that if they did their best, and pupils and their parents more or less co-operated, then the end result would be most pupils making their way in society successfully. In fact, very few children managed the transition into the outside world; many went straight from school into sheltered work and accommodation, and those suffering from degenerative diseases often died. (This is discussed in more detail in Chapter 9.)

In how many ordinary schools are the hopes and ideals that staff had as trainees met? How many children fulfil their own and our expectations? One way of reducing the pain arising from this disappointment is to alter the primary task, which is subtly modified to, for example, 'life skills', or to passing standard examinations which implicitly suggest the child is now equipped for life. A move to something more achievable is sometimes determined by the difficulty in reaching the original goals. And in subtly changing the goals, we lose the opportunity of assessing whether the goals are realistic or whether our approach to them needs to be altered. In other words, our falling into unconscious defensive manoeuvres interferes with our capacity to review the task and to adjust the system appropriately. From an insider's point of view, this process is often very difficult to detect.

DEFENSIVE STRUCTURES IN PUBLIC SECTOR ORGANIZATIONS

For the container to have the best chance of containing and metabolizing the anxieties projected into it, it needs to be in a depressive position mode (see

Chapter 1), which means it has a capacity to face both external and psychic reality. For organizations, this requires not only agreement about the primary task of the organization, but also remaining in touch with the nature of the anxieties projected into the container, rather than defensively blocking them out of awareness. In order for a system to work according to these principles, a structured system for dialogue between the various component parts is necessary. This depends on all concerned being in touch with the difficulties of the task, and their relative powerlessness in radically altering the pattern of life and of society.

The present position of many public sector organizations, however, is quite a different one. The new style of management is to give managers more power and to eliminate consultation as 'inefficient'. It has become a top-down model, with dialogue and co-operation between the different sectors seen as old-fashioned, and care staff increasingly excluded from policy- and decision-making. This style of management could be described as 'paranoid-schizoid by choice', fragmenting and splitting up systems instead of promoting collaboration. The splitting up of functions makes it more comfortable for managers to make decisions.

For example, in health care systems, managers are kept at a distance from the clinicians and the patients. The structure thus enables managers psychologically to turn a blind eye to the consequences of their actions. In the short term, this gives an impression of effective change; in the long term, the consequences are disastrous. Meanwhile, the caring that has, so to speak, been 'leeched out' of the management system is precipitated into the carers, who in turn have left their administrative/financial-reality side in the managers. For example, in Britain, the Griffiths Report (1988) on the re-organization of the health services anticipated that a fair proportion of new-style managers would be doctors. As it turns out, less than 5 per cent are, as if financial concerns and concerns about the quality of care cannot be held within one role.

What effect does the existence of this psychic configuration have on doctors? It makes for the creation of a system, strengthened by group and institutional processes, that fosters individual and professional omnipotence, a system in which weakness, doubt and distress are seen as undesirable qualities and where failure can be attributed to uncaring managers and insufficient resources. An integral part of the unconscious social system that is intended to shield us all from death is the requirement that the office-bearers of the system – in this case the doctors – be as powerful as possible. As a result of the widening gap between the management and clinical sectors of the health service, doctors have become more vulnerable than ever to the projection into them of societal fantasies of their omnipotence, caught up in an unconscious social projective system in which the capacity to do heroic things is imputed to them, and they are expected to perform. Yet it is hard for the system and its functioning to be questioned, for both doctors and the public have a vested interest in keeping it in place. Any tampering with the system creates a great deal of anxiety and resistance on all sides.

FACING PSYCHIC REALITY

In a management climate such as this, in which contact between the various component parts is at a minimum, it is easy for doctors to fall into a state of mind believing that much more would be possible in the fight against death if only more money were available. The shared fantasy between doctors, the public and the media seems to be that we could have eternal life, if only there were unlimited health funds.

Within the hospital, too, the staff need to protect themselves from the reality of illness, pain and death. Walk into one or another institution and you will probably be bowled over by the horror of the place. Mention it to a regular member, however, and they will not know what you are talking about. This is not because it does not exist, nor because they are used to it, which would imply a certain benevolent acceptance. What they are expressing is a denial, or a repression, of the substance of your observation. This flight from reality happens gradually and largely unconsciously. In the process of inducting new members, the group unconsciously gives the message, 'This is how we ignore what is going on – pretend along with us, and you will soon be one of us.' It can be called settling down, or it can be called institutionalization. In fact, it is a collusive group denial of the work difficulties.

Another way of protecting oneself against what is unbearable is to organize the work in ways that ward off primitive anxieties, rather than serving to carry out the primary task. A great deal of what goes on is not about dramatic rescue but about having to accept one's relative powerlessness in the presence of pain, decrepitude and death. Staff are ill-prepared for this in their training, and in their work practice there is often no socially sanctioned outlet for their distress (see Chapter 11). This then expresses itself as illness, absenteeism, high staff turnover, low morale, poor time-keeping and so on. In a study on nursing turnover (Menzies 1960), it was found that it was usually the most sensitive nurses, those with the capacity to make the strongest contribution to nursing, who were most likely to leave – perhaps those least willing to join in the institutional systems of denial.

At a seminar, top health service managers were asked to name their own worst personal anxieties. They mentioned death, debilitating illness, divorce, insanity, abandonment, loss of employment and so on. All of these are of course central to the work (and the workers) of the health service and, indeed, all our public sector services. They viewed their task as one of management, and stressed that the only requirement legally laid down was for them to live within their allocated budgets. It clearly was too painful for these managers to be in touch with the needs of the patients and the consequences of their actions, and psychologically more comfortable to focus on budgets – a classic example of splitting used to avoid depressive-position pain. A great deal of the disorganization, time-wasting on and off committees, bureaucracy and the like is a way of avoiding face-to-face contact with patients and their ailments.

Similar processes occur in our other public sector services: it is contact with pain – the clients' pain and our own – that regularly puts us in touch with our feelings, our impotence and the inadequacy of our training and of our professions. Many of our so-called administrative or managerial difficulties are in reality defence mechanisms arising from the difficulty of the work. Furthermore, a system of financial reward for 'effective' management further bolsters this defensive style of functioning. From a psychoanalytic point of view, we then have a system in which the caring depressive-position functioning of managers and management systems is penalized, and the defensive paranoid-schizoid component is rewarded.

IMPLICATIONS FOR MANAGEMENT

Our public sector institutions can usefully be thought of as comprising three sub-sectors: the public and its consumer representatives (patients, pupils and their parents, etc.); the care sub-sector (the staff of the services); and the administrative system (representing government). So far, we have looked at the anxieties that are being defended against. There are, however, other factors at play of an intergroup nature. These take place within the sub-sectors, and have to do with a rivalry between various professional and administrative sub-groups. They have always been there, but in a climate of increased pressure and, therefore, of increased splitting and projective identification, they are obviously exacerbated.

In the caring sector, the result is more strife between the various professional disciplines and heightened competition for resources. With greater rivalry and reduced communication, the situation is often not unlike the chaos found in group relations conferences (see Chapter 4). There it is for study purposes, to learn about the irrational unconscious processes in and between groups. Here, however, it is for real – and permanent. Within the administrative sector, it is also not uncommon to find massive divisions between the various departments. This of course hinders competent management and encourages a technique of playing one group off against the other. In order for any organization to function at its most effective, certain guidelines, based on group relations understanding and sound management principles, need to be laid down.

For *all* members of the organization, be they cleaner or managing director, there is the need for:

- clarity about the task of the organization;
- clarity about the authority structure;
- the opportunity to participate and contribute.

In addition, for those in authority there is a need for:

- psychologically informed management;
- awareness of the risks to the workers;

- openness towards the consumers;
- public accountability.

Clarity about the task of the organization

As an example, cleaning contracts in the health service now go out to tender. It is clear from the contracts that no account is taken of the fact that while cleaners are there to clean, they are cleaning in a hospital with patients bearing anxiety and pain. The human contact is important for both patients and cleaners. Cutting out the commitment to the overall task is done to the detriment of all.

Clarity about the authority structure

Clear lines of authority make for accountability and therefore for the possibility of changing work practices into more appropriate ones. For instance, in the British national health service, the general manager of any district health authority is accountable to the chair of the local health authority. The latter, appointed by the Minister of State for Health, heads a committee of members who represent local interests. The general manager of the district health authority is, however, separately and independently accountable to the regional general manager. No man can serve two masters, yet this is the present position. Small wonder there is talk of abolishing the local health authorities. With them will go the last vestiges of local and patient representation.

The opportunity to participate and contribute

Contract labour does not make for good staff morale or effective organization, first because contract labour does not have institutional allegiance, and, second, because of the ill-will created in the permanent staff. At one stage, more than half the secretaries in the health service were temps because they could get much better salaries that way than as permanent members of staff. An organization run on a *Gastarbeiter* ('guest-worker' – being a euphemism for disenfranchised staff) principle is not a good idea.

Psychologically informed management

This would include awareness of group and social factors that might interfere with the task of the organization. Such awareness can enable managers to take measures to combat anti-task phenomena. For example, most meetings not only do not start on time, but, more surprisingly, do not have a designated ending time. The logical-sounding rationale is that it depends on how much is on the agenda, and how long that will take. It is widely recognized that ninety minutes or so of committee meetings is as long as anyone can maintain useful attention. Yet this time-span is neither scheduled for nor taken account of. Decisions are therefore made on the basis of grinding down, rather than by working through. Decisions made on the basis of out-manoeuvring or wearing down the opposition do not lead to successful management; they are often Pyrrhic victories.

Awareness of the risks to the workers

For any organization to function effectively the managers must take into account the stresses on the staff as a result of the work they are doing. They need also to make adequate provision for dealing with staff distress, and to ask themselves whether seemingly unrelated anti-task phenomena might not be manifestations of this. It is crucial that a climate is created in which the stress of the entire system can be acknowledged openly, with an awareness of the particular risks to the workers from the nature of the particular task they are performing. These will be different for the helping professions – teachers, social workers, prison officers and so on. One can usefully think of pain, anxiety and distress as being as much a part of the atmosphere and as widespread as is coal-dust in a mine. As in the coal mines, so also in the human services, attention needs to be paid to keeping the 'coal-dust' to a minimum, and to detecting its ill-effects as early as possible, before a chronic or terminal illness develops in the worker.

Openness towards the consumers

Given the defensive tendency in the health service to push patients and what they so painfully stand for aside, it is not surprising that patients are, by and large, forgotten. It is much easier to deal with diseased organs than with a person who has a complaint (often a very genuine complaint), and staff often focus on organs to defend themselves against people-contact. Similarly, senior health service managers may think and talk in terms of populations, again as a defence against the pain of thinking about individuals.

Public accountability

It is a moot point whether the public have a right to know via the media that their local hospital has no beds available for emergencies. Administrators are loath to inform the media of relevant local or national issues. If the information does get out, much time and energy is spent trying to trace the source, and much anger is expended in the process. It seems that the authorities are accountable to those further up the line, and that public information and opinion count for very little. However, the authority for running public sector services derives ultimately from the public (the electorate), and accountability to the public needs to be held in mind and built into the system.

CONCLUSION

Looking at the various defensive patterns described – whether between institutions and their environment, or inter-institutional, or interpersonal – we can see how a style of work that is essentially and consistently defensive is bad not only for the work but also for individual workers. To be constantly out of touch with many aspects of psychic reality at work puts individuals at risk of being out of touch with themselves as a result of a combination of work defences and personal vulnerabilities. There is then an increased tendency to take to alcohol, sedatives, sleeping tablets and so on. This brings the further risk of endangering marital and family mental health. The pattern can influence the behaviour of children and their reactions to stress, and therefore perpetuate itself. The chances of developing stress-related diseases are also increased. We therefore come to the end of the road – an unhealthy mind in an unhealthy body in an unhealthy organization.

Groups and institutions accept newcomers and mould them to the institutional ways of doing things, including joining into their particular version of institutional defences. Eventually, the individual to a large extent loses his or her capacity to be detached and to 'see' things from an outside perspective. Yet, to maintain some outside perspective is essential if one is to retain a capacity for critical thought and questioning. Without these, our institutions are doomed to operate more and more on a basis of denial of reality – the reality that they exist to help people cope with pain, unfulfilled hopes, sickness and death. The more this is denied, the less effective the systems become, and the greater the toll on those involved in them.

Chapter 19

Balancing care and control

The supervisory relationship as a focus for promoting organizational health

Christopher Clulow

The conflict between care and control is endemic to organizational life, and particularly the organizational life of agencies that exist to serve people in trouble. Managers need to balance the two, particularly in supervising the work staff do with their clients. Supervision can and should be a major source of support and learning. It is also an essential mechanism for monitoring the quality of a service, for which managers are responsible. Furthermore, since the relationship between staff and clients in helping organizations itself often includes elements of both care and control, the supervisory relationship is likely to mirror fundamental tensions in the organization's task. As such, it can be a key to understanding institutional processes.

Nowhere is this more evident than in the probation service, where the central activity is supervising offenders. I shall therefore begin with a brief discussion of this service, and then examine some of the key dilemmas in the supervision process, as they emerged in the course of a training programme on staff supervision for probation service managers.

STAFF SUPERVISION AND THE PROBATION SERVICE

For the probation service in England and Wales, the balance of care and control has been a long-standing preoccupation. In the late 1970s, the debate was between those espousing 'treatment' and 'justice' models of probation practice. In the late 1980s, the debate continued in terms of the relative emphasis to be given to punishment and social work help. Against a background of financial stringency, overcrowded prisons and pressure to discourage courts from passing custodial sentences, the government became an active player in the debate, proposing changes in the nature of the relationship between probation officer and offender, with officers carrying out more punitive and controlling activities. Many officers protested about being turned into 'screws on wheels', and about the increased level of central direction, which was regarded in some quarters as a threat to professional identity and autonomy.

This snapshot suggests three factors that might influence the intensity, and visibility, of the conflict between care and control: the nature of the work, in this case supervising offenders in the community; the personalities of those engaged in the work, and their assumptions about its purpose; and the pressure for change, which can disturb an established balance.

The first two factors give rise to the subtle interplay between personal and institutional anxieties and defences, which permeates all aspects of organizational life, and is a recurrent theme throughout this book. (See particularly Chapters 2 and 12.) With regard to the probation service, Woodhouse and Pengelly (1991) point out that probation officers have to manage within themselves and their organizations tensions which their clients have manifestly failed to do. These include not only the conflict between care and control, but also that between dependency and autonomy, individuality and conformity.

The third factor, the pressure to change, is often the spur to invoke the help of an outside consultant or training body. So it was that my colleagues and I were invited to run a course on staff supervision for all the senior managers in a probation service, just after a period in which the service had reviewed its objectives and organizational structure with the help of management consultants. A statement of corporate aims, values and strategy was about to be published, outlining an action plan for the coming five years, with emphasis on corporate identity, the accountability of staff to the service, and the need to ensure that what existed on paper was translated into practice. The supervisory function of senior staff was considered to be of key importance in managing how the changes were to be introduced and accomplished.

A CONSULTATIVE APPROACH TO TEACHING SUPERVISION

Several planning meetings were held with the assistant chief officer carrying responsibility for training to establish what had led to the request and the expectations attached to a course on staff supervision. We wanted to take account of the danger, latent in all requests for training, that an unconscious purpose might be to deflect attention away from organizational needs while giving the appearance of attending to them. We also wanted to avoid stepping into the role of outside experts, not least in the wake of the recently completed organizational review and our knowledge that we were being invited in to help with the process of empowering senior staff to take responsibility for implementing change.

In consultation with the service, we designed a training programme that addressed managerial, consultative and technical aspects of the supervisory relationship. Rather than offer formal teaching, we tried to create structures within which some features of organizational life might be replicated for and experienced by the membership and staff so that we might all learn through working at first hand with the issues that arose. We renamed the course a workshop, underlining the exploratory purpose of the event, while taking responsibility for the temporary institution we were creating and the purpose for which it was being set up. Taking our authority from the chief officer, we founded our role on that of the supervisor, translating into the temporary institution our responsibility to manage the primary task of the workshop (to learn about effective staff supervision), to provide participants with opportunities to develop supervisory skills, and to consult to work issues.

At the centre of the event were work study groups and supervision quartets, each of which owed much to group relations training designs pioneered by Bridger (1990) and Miller (1990a). The work study groups provided opportunities for members to discuss issues arising from their supervisory roles, and also to reflect on how the group behaved as it carried out this task. The supervisory quartets (in which one member presented a real work issue to another, while the two remaining members observed) gave participants an experience of supervising and of being supervised by colleagues, and also of benefiting from the observations of colleagues about the supervisory process as it occurred. These two structures, combined with application groups and plenary sessions, generated the material from which members could learn.

We were struck by the negativity about supervision, from both staff and supervisors, in an agency whose primary task was to supervise offenders. Three work-specific issues helped to account for this: ambivalence towards authority, fear of disclosure, and 'problems with triangles'. Each was a recurrent theme during the programme, and each relates to fundamental tensions in the supervisory relationship. While they surfaced in the context of a training event on staff supervision for probation service managers, they have relevance for practitioners, supervisors and managers in many other organizational settings.

AMBIVALENCE TOWARDS AUTHORITY

The word 'supervision' was used to describe the relationship probation officers had with their clients, as well as the relationship seniors had with their officers. Each relationship mirrored the other and, not surprisingly, both could generate ambivalence. To be supervised had connotations of being placed on probation, not being trusted, failing to reach the right standards and so on. Some supervisors disliked the controlling and the nannying images they associated with the supervisory relationship. This made it difficult for them to take up the role with enthusiasm – especially when these images were shared by their staff. The difficulty was more than linguistic. The following vignette illustrates different approaches to the problem of taking authority in a crisis:

Robert, a senior officer, described an incident which had occurred in an office for which he had joint responsibility with Bill, another senior colleague. A full-scale row had broken out between Philip, an officer, and the secretarial staff over a poster displayed in the office. Philip had acted unilaterally to remove the poster, bringing protests from the secretarial staff and an edict from Bill that the poster should be reinstated. Robert was concerned about the way the situation had been handled, and his own feelings of helplessness in bringing about a less didactic solution to the problem.

Attention was drawn to what was being enacted between Philip and the secretarial staff and its meaning for the two senior officers. One senior had acted by edict, arbitrarily reinstating the poster, like the officer who had unilaterally taken it down. Having repudiated the stereotypical image of controlling male authority, the other was left feeling helpless and rather resentful at not having had any influence in the matter, like the secretarial staff. The implications of this incident for the relationship between secretarial and probation staff, the style of office management when both seniors were men, and the relationship between them, provided a fruitful focus of discussion for the programme members.

Equally disliked, but sometimes seen as alternative to control, was an indulgent 'nannying' image of the supervisory relationship, as seen in the following:

> Concern about becoming a nanny was expressed by Michael, a senior officer who was supervising Gillian, a very able new officer. Gillian willingly volunteered to take on lots of new work; she was also prone to bouts of sickness. Michael expressed concern that Gillian was relying on him more than she should, but he was finding it difficult to refuse her demands on his time. In his view, Gillian was competent enough not to need the amount of support she claimed from him.

Two possibilities for understanding what was going on are suggested by this vignette. First, Gillian might appropriately be drawing attention to aspects of her work with which she needed help but which she did not feel entitled to ask for directly. In other words, the fear of nannying might be shared by supervisor and supervisee. Second, the difficulty Michael, the senior, was experiencing in saying 'no' to claims on his time could well be the kind of problem a new, bright officer might be having in protecting herself from becoming over-committed with her clients.

Both possibilities assume that communication between supervisee and supervisor occurs not only in what is said, but also in what is enacted between them. Moreover, they assume that unconscious enactments in the supervisory relationship can reflect work dilemmas between supervisee and client that have yet to be distilled into conscious awareness. This dynamic link between practice and supervisory contexts has been described as the 'reflection process' (Mattinson 1975). Recognizing the link suggests that it is not only what the supervisor says, but also how he or she manages the experience, which is of importance to the supervisee. Managing the supervisory boundary, as well as making links between what is happening in supervision and work contexts, provides opportunities for translating enactments into insight.

The controller/nanny dilemma was echoed in other discussions, notably about writing reports evaluating supervisees' performance. On the staff supervision programme, it was evident that preparing evaluation reports on colleagues could be difficult for probation managers, who felt themselves to be on the horns of the control/care dilemma. No one wanted to become either judges or caseworkers in relation to colleagues, but sometimes it felt as if those were the only options available. The result could be either bland and depersonalized evaluation exercises, or reports which wallowed in the idiosyncrasies of staff personalities. (Chapter 21 considers some of the connections between evaluation, practice and institutional anxieties in more detail.)

The process of developing an institutional expectation of committed supervisory practice could not avoid engaging with ambivalent feelings in the staff towards authority. This ambivalence, in turn, could not be disconnected from the nature of probation work and the client group of the service. Indeed, unresolved conflicts about authority and control often figure in the choice of this profession (see Chapter 12). Sometimes, ambivalence towards authority resulted in lip-service being paid to the need for supervision, without any real conviction about making an active commitment to it.

FEAR OF DISCLOSURE

Ambivalence towards those in authority, as well as being a learned attitude, can be founded on a profound mistrust of other people. In their study of a social services department, Mattinson and Sinclair (1979) found that clients frequently managed their distrust of social workers, who carried many negative attributions from clients, by dividing them into 'suckers' and 'bastards'. The split was one way of managing the conflicting desire for and fear of those who offered help, especially those in positions of authority. Self-disclosure in these circumstances was a risky business, calling for high levels of trust, and a capacity to place oneself in a vulnerable position.

Our training event to address supervisory practice, particularly since it was run 'in-house', inevitably raised similar anxieties associated with disclosure, and for some of the same reasons. The requirement that people talk about and demonstrate how they actually work generates fears of criticism, loss of credibility, or worse. This anxiety about dangerous consequences following on self-disclosure mirrors clients' reservations, both in social work and in probation work, and is at least partly a projective identification (see Chapter 5).

The issue of confidentiality was raised early on. A member asked what protection there would be for officers who were not on the programme, who were identifiable, and whose work would be discussed in the context of examining supervisory practice. This proper concern to respect confidences was responded to by our restating the task of the programme, which was to address the supervisory practice of members and not the performance of the staff they supervised. In the discussion which followed, it became apparent that some of the concern being expressed was as much about how safe it was for senior officers to disclose details of their practice to each other as it was about preserving the confidences of supervisees. Could colleagues be trusted not to abuse confidences? Was it safe to reveal one's incompetence? What would happen if you invited confidences which then required action to be taken? How were feelings to be managed? These questions were pivotal for supervisory practice both within the organization and in relation to offenders.

Concerns frequently turned out to be related more to disclosing feelings, such as anxieties about competence, than to confessing acts, omissions or 'offences' of one kind or another. Preserving a sense of credibility, 'professional reputation' being the equivalent of 'street cred' for clients, was important for staff, and fear of losing face could be infectious. Clients who failed to report for appointments could cause officers to doubt their own credibility. Supervisees who were reluctant to be supervised could cause supervisors to question whether they had anything to offer. (Both, of course, also raised suspicions as to what the non-attender might be hiding.) Consciously or unconsciously, reciprocal arrangements could operate between supervisors and officers to support each other's defences. An officer who thought it unwise or unsafe to disclose failings, real or imagined, might conceal these from the supervisor. A supervisor who felt unsure in role might use pressures on time to collude with the officer in ensuring they never met, or, if they did, that the meeting prevented the real concerns of either side being talked about. One senior described how he buried himself in paperwork as a means of dealing with anxiety about an increasingly pressured workload.

The fear that to talk about work problems may result in being regarded as a failure can increase the risk of producing a casualty – either directly in the person concerned, or by passing anxiety down the line to others. Acknowledging and addressing this fear is a priority in the supervisory relationship if disclosure is to be encouraged. Legitimizing the feelings of supervisors and supervisees is a precondition to attending to what they might mean, and how they might be deployed to the benefit of clients.

PROBLEMS WITH TRIANGLES

Supervision is a three-way process involving the client, the supervisee and the supervisor. In the context of probation work, the offender is the client, who receives a service from a probation officer, on behalf of the agency. (The service has other clients, notably the courts; the question of who the principal client is touches on the care and control debate referred to earlier.) The senior officer, through supervision, represents the agency which regulates the nature and degree of transactions between officer and client. Difficulties can be anticipated if the client is forgotten in considering the officer's predicament; if the officer is lost in considering the client's predicament; or if the senior (the agency) is lost in the officer's work with clients. Maintaining the *triangle of supervision* (Mattinson 1981), by always holding all three in mind, is essential if an organization is to carry out its primary task effectively.

Managing threesomes is notoriously difficult. The attempt to ignore or forget one part of a triangle can be understood as a defence against anxiety about being either helpless or destructive in relation to the whole. For a senior to intrude upon the relationship between officer and client by suggesting the need for supervision, for example, or for a client to intrude upon the relationship between officer and senior by lodging a complaint, or for an officer to intrude upon the senior's special relationship with the chief officer, perhaps by going over the senior's head, raises anxieties about competence and fantasies about damage.

It is not uncommon for supervisors to experience the 'emperor's new clothes' syndrome (Dearnley 1985), fearing that their ignorance will be transparent if they engage in supervision. Nor is it uncommon for supervisors to feel they are being destructive if, spotting practice which is potentially damaging to client or agency, they act to intervene. Through feeling either inept or persecutory, the fantasy can arise that to intrude upon the pairings of others will result in disaster. When both the supervisor and supervisee have anxieties about this, they may collude in avoiding supervision sessions, despite the cost both to the supervisee and to the service.

A senior, Alan, questioned his competence to supervise the work of Daniel, a specialist officer in his team. He did not wish to intrude on the work, but felt he should know what was going on. However, he was diffident about pressing for supervision sessions, which Daniel had been cancelling, because he felt he knew very little about the specialist aspects of Daniel's work, and that Daniel might therefore gain little from them.

Implicit here is the notion that supervisors are technical experts who need always to possess more knowledge than their supervisees if they are to feel competent in role. Not only does this deny the usefulness of 'lay' questions, and lose sight of the importance of maintaining a space for reflection and the value of the senior's overall experience, but it also denies the accountability of individual officers to the agency for the services they deliver to clients.

An alternative to feeling shut out by the practitioner/client relationship is for supervisors to over-identify with supervisees, at a cost to their management role.

Tara, a senior officer, complained that the good relationship she had with a supervisee, Damien, was being marred by a policy decision that required Damien to move from his 'beloved patch' to overseeing community service work. Tara felt as if she were 'piggy in the middle' of a fight between Damien and headquarters staff. She also felt she was being pushed into a surveillance role with Damien, just as he felt he was in relation to his new clients. The marked change from their previously warm and friendly relationship was experienced by Tara as a personal loss, and temporarily disabled her from supporting Damien in his transition into a new role and type of work.

CONCLUSION

In maintaining the triangle of supervision, the capacity to develop a 'third ear' is an indispensable asset. Without an ability to stand back and decode communications contained in the supervisory relationship, both supervisors and supervisees in an organization may be at risk of replicating the enactments of their clients rather than attending to them and containing them (see Chapter 7). An integrated, internalized model of the supervisory function comes about through oneself having had a good-enough experience of being supervised. The capacity to stand both inside and outside a situation, to hear what is said as carrying symbolic as well as literal meaning, can be transmitted from supervisor to supervisee. If that happens, there is reason to believe it can also be transferred from practitioner to client, reducing their need to communicate through enactments which can bring them into conflict with the community.

In this chapter, we have seen how the supervisory relationship can act as a receiver of unconscious communications which bear upon the nature of work-specific anxieties, and also some of the individual-cum-organizational defences that are deployed to manage them. In this relationship more than in any other organizational relationship, there are opportunities to decode these messages and attend to the anxieties they convey, thereby contributing to understanding, learning and good practice. Precisely because of its location and potential, the state and quality of the provision made for supervision can be taken as a key indicator of organizational health in the human services. Ambivalence about authority relationships, fear of disclosure, and problems with triangles are not the sole province of agencies that deal with offenders: they permeate every organization. How they are managed affects not only the current health of organizations and those working within them, but also the degree of space there is for change and development in the future. This, then, is a matter of survival. Organizations that are unable to adapt to a changing environment, and to be sensitive to signals from outside and within, are unlikely to survive for long.

Chapter 20

Conflict and collaboration

Managing intergroup relations

Vega Zagier Roberts

Throughout the human services, complex tasks require that members of different groups – whether departments within a single organization, or different agencies dealing with the same clients – work together. There is much talk about the need for better co-ordination, for collaboration, for teamwork. Yet services continue to be fragmented, intergroup rivalry and conflict are rife, and attempts to address these difficulties are met more often with frustration and failure than with success.

INTERAGENCY RELATIONS

As services for people with severe and long-term physical, mental or emotional difficulties and other high-need clients have increasingly moved out of the large institutions and into the community, the various activities needed for their care have been taken up by different agencies. Thus, an individual may be involved with his or her doctor, the local community psychiatric nurse team, a social services day centre, the housing department, a voluntary sector social club and a host of other helping professionals. Each agency deals with its own 'bit' of the client, and new problems are likely to lead to further referrals to yet other agencies.

Without co-ordination, gaps and/or overlap in services are likely to occur. As the client passes from one agency to another, each can blame the others for any difficulties.

New Start was a voluntary sector organization set up as an alternative way of meeting the multiple needs of people leaving long-stay psychiatric hospitals to live in the community. Its aim was to provide the whole range of support services from within one team. Its members were surprised to get very few referrals directly from the local psychiatric hospital; instead, most of the referrals came from the housing department, which wanted help in reducing the high rate of breakdown and emergency re-admission to hospital among the people they placed in bed-and-breakfast accommodation. However, the New Start team were turning down nearly two-thirds of the referrals from housing as not meeting their carefully thought-out criteria for which kinds of clients they felt able to take on. Each time they turned one down, they re-stated these criteria, complaining among themselves that, yet again, statutory agencies were disorganized and trying to use New Start as a dumping ground.

The housing department placement team suggested the two agencies do joint assessments of patients about to leave hospital, in order to determine who was most suitable for available accommodation. The New Start team refused, saying this was not their job. Meanwhile, patients referred to and placed by the housing department continued to break down and return to hospital in large numbers; referrals from housing to New Start were being turned down; and there were hardly any referrals from the hospital to New Start.

I had been asked to consult to the New Start team to assist them in developing their service. As we began discussing their aims, team members talked about their shared view of statutory agencies as being paternalistic and condescending towards clients, abusing their power and fostering helplessness. They were deeply committed to empowering their clients and to fighting abusive practices on their behalf. As a new and small organization, they felt vulnerable to being exploited by the larger and more powerful agencies, as did their clients, with whom they identified consciously and unconsciously.

They were also anxious to prove they were capable of supporting their clients effectively. To this end, they had evolved their criteria and defined their role as picking up where the others left off. They were very worried about those people who were 'getting lost' – the potential clients who were not being referred to them by the hospital and other community agencies – but were ambivalent about pursuing these lest they be overwhelmed by demands they could not meet.

Their refusal to participate in joint assessments could now be understood as having several sources. The most conscious was a fear of being exploited, which led to their holding firmly to the boundaries they had set on their task. Less conscious was anxiety about having power over clients if they got involved in making decisions for them, when they saw their role as protecting clients from others' abuse of power. Finally, there was anxiety about feeling more responsibility for clients' breakdown. Although they felt uncomfortable about the referrals they were turning down, as well as about the ones they were not getting, they could argue the blame did not lie with them. By disavowing power and locating it elsewhere, they reduced their guilt and sense of responsibility, but at the cost of feeling more helpless and vulnerable than was necessary.

Over time, the New Start team came to see that the problem of clients' successful transition from the psychiatric hospital into the community was one shared by all three agencies, and took the initiative in setting up what proved to be a very successful joint assessment scheme to replace the former sequential model of interagency referral.

INTERDEPARTMENTAL RELATIONS

The example above illustrates some of the problems which arise when different agencies are working with the same clients. Similar processes occur within a single organization where different departments are contributing to the overall task. For example, in a hospital, a number of departments – medical, nursing, housekeeping, volunteer services and others – are likely to be contributing to the well-being of the patients, but often with little co-ordination.

The rivalry, conflict and mutual disparagement between nursing and the other departments at Shady Glen have already been described with regard to the need to define a task which was both meaningful and feasible, and to which staff from all departments could feel committed (see Chapter 8). The consultation, initially offered separately to nurses and non-nurses, eventually brought the two groups together into a joint working party, to prepare a proposal for improving patients' quality of life. This proposal contained many ideas for changes in working practices, but a central recommendation was to alter the boundaries of the task-systems.

Originally, nurses of all grades, including students, were based full-time on a particular ward; everyone else was firmly based in their respective departments, visiting the wards or removing patients from them to provide their particular input. For instance, occupational therapists (OTs) ran cooking and art groups in the occupational therapy department, and waited there for patients to be brought from the ward. When the ward was short-staffed, there was often no one available to escort the patients to their groups, and patients would therefore not arrive that day. The OTs also ran current events and reality-orientation groups on the wards. Very often, they would arrive to find none of the patients ready, as they were still being bathed and toileted. After waiting for a while, the therapist would leave, angry at her efforts being sabotaged by the nurses. Physiotherapists and speech therapists encountered similar problems. The nurses, meanwhile, felt resentful that they were left to do all the heavy physical work of lifting and moving patients to clean and dress them while the OTs could 'swan in and out'.

Our recommendation was to enlarge the boundary of the ward team to include everyone contributing directly to the re-defined primary task. This had the effect of reducing rivalry, conflict and unconscious sabotage of each other's work across departments. It also made possible some creative re-thinking about how various activities related to the overall task. For instance, when an OT arrived to lead her current events discussion group but found that what was most urgently needed was another pair of hands to help with toileting bed-bound patients, she could help out with this 'nursing' task. This might then free some nurses to join the group activity, rather than cancelling it because they were 'too busy'. Finally, patients' failure to improve was now a shared problem, rather than something to be blamed on someone else.

INTERGROUP RELATIONS IN MULTIDISCIPLINARY TEAMS

At Shady Glen, the glaring drawbacks of splitting up patient care among different departments led to our proposing a new task-system which would bring staff from all disciplines together. However, in many wards, day centres and community health care settings, such a system already exists: the multidisciplinary team. The hope is that these teams will plan and provide an effective co-ordinated service, since they have all the specialist skills needed under one roof. Instead, they often duplicate the same interdisciplinary fights and rivalry, or failure to co-ordinate their various activities, as have been described above. Other teams, determined not to repeat these mistakes, encounter a new set of problems.

> Bradley Lane, a community mental health centre staffed by a multidisciplinary team, had been plagued for years by bitter splits and fights between the various professional groups. The psychologists, psychiatrists and social workers spent most of their time offering individual counselling and therapy to the centre's least-disabled patients. Nurses and occupational therapists ran group activities for the more chronic patients, and often resented being allocated this 'low-status' work.
>
> This changed when the newly created post of team manager was filled by an enthusiastic young psychiatric nurse, Anna. Anna had been working in a therapeutic community and brought with her an inspiring conviction of the value of team cohesiveness. Over the next few months, the team spent a great deal of time reviewing their working practices, including two days of team-building with an external consultant. On the basis of this, they decided to work generically: all of the centre's activities would be shared across all the disciplines. Clients would be assessed by the whole team and then assigned to whomever had a vacancy, regardless of their particular profession. The groups offered to patients gradually shifted towards an emphasis on insight and change, rather than on training in social skills.
>
> At an annual review, the team psychologist noted there was a gap in their service, in that clients with full-time jobs found it difficult to attend any of the available treatment programmes, and suggested that an evening therapy group might be useful. As she could not run such a group herself because of family commitments, one of the nurses, Sharon, offered to do it. She spent many weeks planning and publicizing the group, with help from colleagues. When Sharon went to her line manager in the nursing department to ask for time off during the day in lieu of working one evening each week, he refused, saying the nurses' priority should be to expand the day programme for the most chronic patients; therapy should be offered either by psychologists or by social workers.
>
> The whole team was shocked and outraged at having their plan overturned. After a burst of protest, including a letter to a more senior manager which had no effect, there was a steep decline in morale and three people left the team

soon after. Even a full year later, they were still refusing to initiate any new projects on the basis that 'they won't let us do anything anyway'.

INTERGROUP COLLABORATION

The word 'collaboration' is often used interchangeably with 'cooperation', to denote harmonious working together. In this chapter, however, it refers to the particular situation where a group of people come to work together *because of* their membership in other groups or institutions whose tasks overlap. For example, staff from New Start, the local psychiatric hospital and the housing department had been assessing many of the same clients separately and sequentially. Through forming a joint assessment team, they were able to carry out this task more effectively. At Shady Glen, members of previously competing departments came together in an enlarged ward-team in order to provide a better service to their patients. At Bradley Lane, an intergroup system was already in place: the multidisciplinary team.

In Figure 20.1, the outer boxes represent the original or 'home' groups – the different agencies, departments or disciplines – and the inner box represents the intergroup system or 'collaborative' group.[1] The overlapping boundaries indicate that members of the collaborative group continue to be members of their home-groups, which are task-systems in their own right for activities which do not require intergroup collaboration.

The problem of dual membership

Every member of a group is likely to be also a member of other groups. These may be outside groups – trade union, church, family – or sub-groups within a team, such as the old-timers and newcomers, or men and women. To this extent, every group is actually an intergroup, with intergroup relations which need to be managed. Each group membership carries a greater or lesser degree of *sentience* or emotional significance (Miller and Rice 1967). From this stem loyalty and commitment to the group's aims. Inevitably, individuals with membership in

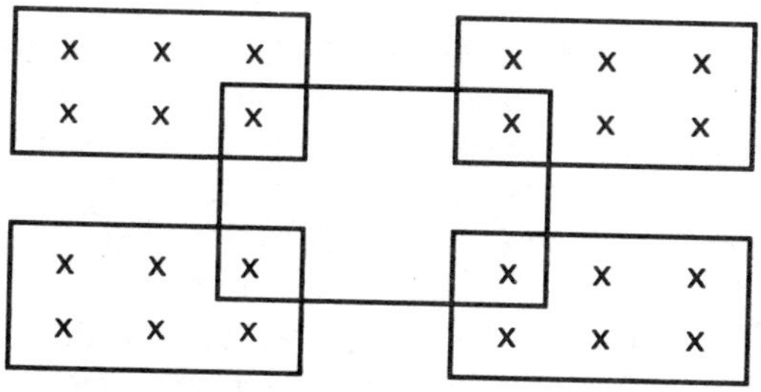

Figure 20.1 An intergroup system

1 The figures in this chapter are adapted from Miller and Rice (1967).

more than one group will sometimes have trouble with conflicting demands from the various groups they belong to, and their dominant group sentience may shift over time.

Most such multiple memberships are incidental as far as the work group is concerned. The teacher who is a member of the Labour Party, the nurse who is a Catholic, the social worker who is a mother, all bring with them these other 'memberships', which affect how they do their jobs; but when they leave their jobs, they will be replaced by others with different affiliations. In other situations, one's membership in another group is part of the reason for being selected for a post: a team may wish to have a certain number of black workers, or of women, or of residential social workers, in order to carry out its task better. Finally, there are situations where the home-group membership is the main reason for being included in the collaborative group. For example, in the joint assessment team set up by New Start, each member was there as a representative of his or her home-agency. Many committees, such as joint planning committees, are representatives' groups, made up of people sent to protect the home-agency's interests, as well as to contribute to the joint project. In this case, it becomes crucial, both for the individual and for the group, to manage this dual membership.

When a new collaborative group first comes together, its members are likely to identify themselves predominantly in terms of their home-agencies. This may well make it very difficult for the collaborative group to work effectively. Members with loyalty to different home-groups are likely to be competitive, and the collaborative group may fragment into fighting factions. Often, however, members gradually invest more and more in the collaborative group over a period of time, as its task takes on meaning and importance. The group builds up a shared value-system, as well as personal relationships among members. As their sentience shifts, they may become more committed to the aims of the collaborative group than to those of their home-groups, even to the extent of 'forgetting' their original membership. This is particularly likely to happen when the collaborative work is done outside or away from the home-group, or when workers spend much more time in the collaborative group than in the home-group (see Figure 20.2). Here, the collaborative group has become quasi-autonomous, a closed system whose members have lost touch with their dual membership and its linking function, on which their effectiveness depends.

For instance, in the case of Bradley Lane, if Sharon had held in mind that she was part of the nursing department, as well as of the multidisciplinary team at the centre, she might have kept her manager more in touch with her plans for an evening group, and even have gained his support. Alternatively, she might have been more aware of the nursing department's priorities and view of the nurse's role, and therefore not have put herself forward to run the group. The idea might then have been taken on successfully by another team member, whose department regarded it as an appropriate use of their time.

It was, of course, not only Sharon who failed to hold her dual membership in

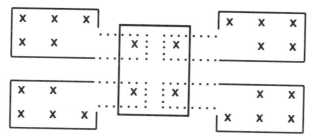

Figure 20.2 Changing sentience in an intergroup system

mind. The whole team was anxious to maintain their hard-won cohesiveness, hence the decision to work generically. In the process, they had obliterated differences in skills, training and experience among members, as well as 'forgetting' their home-group membership. One obvious cost was the damage caused by the 'veto' – damage to team morale, and the loss to the clients who might have benefited from an evening group. There was also a chronic 'running cost' in that the denial of differences disabled individual members from using their special skills – even the very ones for which they were hired – lest they arouse envy and re-kindle the old competitiveness and fights. Anything which could not be done equally well by everyone could not be done by anyone. The team – and their clients – were deprived of the richness of specialized contributions to the overall task, and individual staff members lost a major source of job satisfaction.

The effectiveness of collaborative groups depends largely on their members' ability to manage their dual memberships. Excessive commitment to either membership at the expense of the other will inevitably compromise task performance, and lead to problematic intergroup relations.

The problem of dual management

Often, collaborative groups have no formal management. The members are managed from their home-groups and come together as 'equals', that is, with no one within the group having authority over anyone else. Thus, when Bradley Lane opened, there was no manager of the centre, only a 'co-ordinator' to order supplies and plan rotas; staff were managed from their respective departments. This contributed to the difficulty resolving disagreements in the team, since it was up to individuals (and their department managers) whether or not they abided by any decisions the team made. Concurrently with the team-building work, the health authority was restructured, and as part of this a team manager post was created at the centre. Each member then had two line managers. This is often the case in collaborative groups (see system E in Figure 20.3).

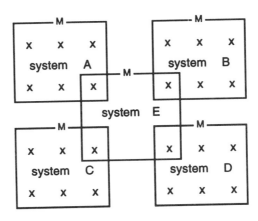

Figure 20.3 Dual management in intergroup systems

Where task-systems have overlapping boundaries, and their members are part of two management systems, the question of who is managing what – where authority is located – can become critical. Many collaborative enterprises founder because this question is not adequately addressed. For example, at Bradley Lane, the team assumed they had sufficient authority to decide to run an evening group and to decide among themselves who would do it. This proved not to be the case. But the nurse manager's refusing permission to a nurse was taken as a veto of the whole plan, which was abandoned without any consideration of the possibility that someone from another discipline might take on the project. Everyone felt equally injured by the decision, which might otherwise have been a blow only to Sharon. The situation was further complicated by the ambiguity about the nurse manager's authority. Was his refusal based only on priorities within the nursing department, or was it a message from senior management levels about what the team's priorities should be? This was never clarified, and contributed to the shared fantasy in the team that 'they won't let us do anything anyway.' With this, they projected all their authority, and lapsed into resentful apathy.

Similar difficulties arose at Shady Glen. The proposed new ward-team boundary, which included staff from other disciplines, needed to be managed by someone with sufficient authority to make decisions about patient care. The most obvious person to take on this management role was the ward sister. Previously, all non-nurses had been managed from their own department; for the enlarged ward-team to be managed as a system, some of the authority previously located in these departments would have to be relinquished to the most senior nurse on the ward.

This did not happen. While our report was still being discussed among senior managers, the specialist therapists became more involved on the ward, as described above, and attended ward meetings, where some innovative ideas about changing working practices and personalizing the patients' living space were developed. In the end, however, the necessary authority was not delegated to the ward sister, and as a result, fundamental change to the continuing-care system as a whole – and to the patients' and staff's quality of life – did not take place. The failure to implement the report came as a devastating disappointment to the therapists. The nurses' response was more along the lines of, 'Well, what do you expect?' – the beginning of a move back to the old splits along the fault-lines between the different disciplines.

SUCCESSFUL COLLABORATION

Of the three collaborative groups described in this chapter, the most successful was the joint assessment team convened by New Start. Its task was the most circumscribed of the three, and could readily be seen to be in the interests of each of the home-agencies, as well as of the clients. Furthermore, its work represented a relatively small proportion of the total work of its members, so that managing dual membership was not as difficult as in the other two examples. Furthermore, delegating sufficient authority to the collaborative system did not pose a threat to the home-groups. All three agencies were needed for the task, and each had a specific contribution to make. Management of the intergroup system also proved fairly straightforward. Since the joint assessment team met to discuss how to plan for clients *before* they were discharged from hospital, the clients were at that point still the responsibility of the hospital. The meetings therefore took place there, so that the staff from New Start and the housing department could meet with patients before discussing them. The meetings were chaired by whichever nurse knew most about the patient's current situation and past history. In summary:

- The task of the collaborative group was clear and feasible.
- It did not conflict with the aims and priorities of the home-groups.
- It was important enough to its members for them to invest sufficient commitment to do the work.
- It was important enough to the home-groups involved for them to allocate sufficient resources (mainly staff time) and delegate sufficient authority to the collaborative group for the task in hand.
- Membership of the collaborative group related to its task, so that each person felt that he or she had and was seen to have a specific and needed contribution to make.
- The group created a management system based on what was needed to achieve the joint task.

To put it another way, the group and intergroup relations could be managed because there was sufficient sanctioning of authority, both from within the group ('below') and from outside ('above') (see Chapter 4). These are essential conditions for successful collaboration, and one or more was missing in the other cases described.

CONCLUSION

Working well together – whether between individuals or across groups and organizations – is generally considered a good thing and, as such, to be pursued without question. Yet before effective systems for working together can be set up or adequately managed, there are basic questions which need to be asked. The first is whether there is a task which requires collaboration. If so, who needs to work with whom in order to carry out this task? And finally, what authority will they need to have, and how (and by whom) are they to be managed? Without adequate attention to these questions, there is every likelihood either of too much togetherness or too little. Too much can give rise to numerous large meetings which feel pointless but take up a lot of time; or to the pursuit of 'cohesiveness' at the expense of individual initiative and the on-task exercise of specific competencies. Too little can result in insufficient co-ordination of related activities, and chronic strife. On the other hand, well-managed intergroup relations, including relations among sub-groups within a single team, can do much to improve both morale and effectiveness in the human services.

Chapter 21

Evaluation

Organizations learning from experience

Rob Leiper

If experience is the food of learning, then evaluation is the digestive process. To benefit from experience, it is essential to make something of it. Any open system – whether an animal, a person or an organization – must learn from its experience of its environment in order to survive. Evaluation, therefore, is one of the essentials of living. It involves taking in feedback from the environment, reflecting on it, considering its value, and making judgements about its implications for future choices. The capacity to adapt depends on a willingness and ability to undertake this process, implicitly or explicitly.

Times have changed in the public services. To an unprecedented extent, professionals are being asked explicitly to evaluate the work they do. This expectation comes in various guises: quality assurance, clinical audit, consumer feedback or individual performance review. All of these have as their core the evaluation of the service being provided. Like any change, it is the occasion of resentment and cynicism, as well as of misplaced enthusiasm. However, it is a reality which must be faced. Formal systems of evaluation have uses as well as abuses, but they can have a constructive place in the life of organizations working towards healthier functioning.

It has been said that a professional qualification is a licence to make mistakes. While double-edged, the positive meaning of this is that a part of being professional is shouldering the burden of responsibility for appraising one's own work and reflecting upon it in order to learn the lessons that experience has to teach. Various opportunities and systems have always existed to enable the discharge of this responsibility, such as supervision, case conferences and further professional development. Good practice has always placed a premium on continuous learning and mutual work review. However, such systems have been subject to evasion and abuse, and public suspicion of professionals has grown, along with a call for greater openness of information, more responsiveness to the concerns of consumers and clearer public accountability. This change in the climate within which services operate has created a demand for formal structures of monitoring and evaluation, which has grown in parallel with apprehensions inside as well as outside public services about instances of poor quality and frank abuse in institutional care.

Furthermore, the search for 'efficiency' has become a dominating pre-occupation. The contract culture requires the clearer specification of service inputs, but it also devolves responsibility for the ways in which the service operates. At the same time, community care policies are creating a dispersed pattern of service provision, raising concerns about the overall control and co-ordination of services. Thus, the normal need for continuing self-appraisal by services must now contend with an increase in the organizational pressures which demand it. The response to these realities can be a relatively open and adaptive one which incorporates new expectations into an acknowledged need to learn. However, the speed of change, the tone of criticism and financial constraints are likely to create a more paranoid and defensive stance. Is it possible to conduct evaluations without producing self-defeating reactions, and which aid organizations in a process of growth?

SETTING UP AN EVALUATION SYSTEM

Southside, a community mental health centre serving a large area within a major city, had been opened eighteen months previously, partly with money freed up by the run-down of a large psychiatric hospital. It was staffed by a multiprofessional team of six, whose brief was 'to provide a comprehensive psychiatric service to the local community, including preventing admissions to hospital and reducing the risk of future serious mental health difficulties'. Along with other community units, Southside was directed to institute 'locally appropriate quality assurance systems' as part of the service's application for status as a self-governing trust. I was asked, as an external consultant, to help units to do this.

Southside faced a number of problematic issues in its work, notably problems about which clients it should devote most time to: the steadily growing numbers of referrals from GPs, which fitted with its ethos of 'prevention'; or those clients with a history of past psychiatric admissions, the group who had received services from the closing hospital from which funding had come. It appeared that services had drifted in the direction of catering for less needy, and so less challenging, clients. Southside lacked a sufficiently clear conception of its basic task; as a result, there was little to aid the centre in deciding upon its priorities.

Behind these competing goals were disagreements between different professions about core values and appropriate models of care. There was also disagreement over the consultant psychiatrist's right to impose his view concerning the appropriate treatment programme for specific clients, and even about the team co-ordinator's authority to allocate and monitor cases. In the face of these disagreements, there was a growing tendency for workers to operate as individuals with their clients, coming together as a team only to discuss their own internal arrangements. Planning and reviewing the care of individual clients had become perfunctory.

Given all these uncertainties and conflicts, the directive to set up a quality assurance system was experienced as an additional stress, adding to the staff's considerable, but only half-acknowledged, difficulties. When I was called upon to consult to Southside to help develop such a system, I wished to help them to find an approach which might move the whole service forward in a constructive direction.

The process of evaluation has a common underlying structure, but it can be realized in many different forms which give rise to numerous variants. The basic ground plan involves clarifying the purpose of a service, translating this into specific means and goals, observing achievements and comparing the results with what had been intended. This process of evaluation needs to be action-oriented, leading to problem-solving and change; and it should be ongoing, a continuous process of learning by reviewing the results of action. There exists an array of choices for translating this basic ground plan into action. The options for what data are relevant and how they might be obtained are many, and each has its own strengths and weaknesses.

Any evaluative process is based on the established aims of a service; it therefore provides an impetus for an organization to clarify its values and its goals, and to do so in terms which are specific enough to be operationalized and monitored. Management and staff are thus pushed to agree priorities, and in doing so are likely to be faced with deciding whether these are attainable within the resources available. Southside would need not only to face some difficult decisions about its goals, but also to make realistic choices about the evaluation work itself, which could not cover everything. It is essential to focus on some aspects of the work rather than to be swamped by trying to change everything at once. The starting-point should be to discover which questions it would be most helpful to answer. However, any such internal review runs a danger of collusion by staff to bypass or cover over the most painful or problematic issues which they face. To avoid this and maintain some external objectivity, it can be very useful to have an external consultant to the review process.

I met with the staff team as a group, and with some individually, to gain an understanding of the centre, and to present a case for quality assurance to them. Their reluctance and suspicion were apparent, but they also responded to the opportunity to represent their work to a management whom they experienced as distant and indifferent. A further meeting with the area manager was arranged, and I supported him in seeing the team as having the authority to review their own work, while expecting them to inform him of the problems they were experiencing. This produced a more positive, even excited, tone, but one which was without direction.

I proposed a system of structured problem-solving meetings over a limited period of time, using pre-existing schedules which addressed typical problem areas in mental health services, and facilitated by me. The team wished to review virtually all aspects of their service immediately, and yet times for the meetings were difficult to arrange. I had to push the staff group on both these issues; both involved setting priorities, which of course reflected one of their main problems. They had to give something up – both other commitments to make space for the review, and the desire to evaluate comprehensively by making a choice of focus.

Hesitantly, the team agreed to start by looking at the aims of the centre and their philosophy of care, and to assemble some information from their current records about the work being done, particularly about which clients were receiving what services. We would then look at the team's ways of working together, both on cases and on policies, with information collected initially from individuals by questionnaire, followed by feedback to the group. The occupational therapist volunteered to join me in meeting a group of the long-term clients to discuss their views of the service. Four team meetings of two hours were set for this stage of the work.

The process of review, as might be expected, did not proceed entirely smoothly. After initial enthusiasm, frustration set in and attendance at review meetings became problematic. Some staff looked bored and disengaged with the detailed discussions of definitions of priority groups. This was only eased when I allowed them to express their frustration directly with me and the review as well as with management. This freed them to acknowledge their overload and the neglect of the long-term clients as problems they would have to tackle.

By the third meeting, they were in a position to bring to the surface, with the help of the questionnaires, some of the interpersonal tensions in the team, and to look at how poor their meetings had become. By the fourth meeting, there was reluctance to conclude the review, which was seen as difficult but supportive.

A tension in all systems of quality assurance exists between, on the one hand, attaining a degree of objectivity so that taken-for-granted working practices may be questioned critically; and, on the other, creating a sense of ownership of the process and its results by the staff who will have to take action on the conclusions reached. It is never easy to get the balance right. There is a constant temptation for staff to split off the responsibility for self-criticism and then see it as malicious or ill-informed. Whatever self-reviewing system is adopted must be internalized by the service and become a part of its functioning, supporting a culture of learning from experience. In an overburdened setting like Southside, this is particularly problematic.

RESISTANCE TO EVALUATION

Any quality assurance process has a number of possible functions. On the one hand, the goal of quality control is to prevent unacceptably poor practice and maintain agreed standards. A more ambitious goal is quality enhancement, where the aim is to raise standards of practice overall, to educate about good practice, and to anticipate and prevent problems arising in the future. Difficulties within health and welfare service systems are usually considerable, and helping them to improve seems to be a more significant aim. The movement towards an organizational culture of accountability therefore opens a new door by which services can enter upon a process of growth and development.

However, the element of accountability underlines a further function: the communication of information about the service to other parts of the system to enable co-ordination of resources and permit other parties with a stake in the service (users, carers, referrers and purchasers) to be reassured about its quality. This complex network of interests with which services are now required to interact creates additional strains of its own. Individual service settings, like Southside, are likely to feel more exposed to demands from and scrutiny by a wider range of stakeholders in their service than they have ever been accustomed to in the past. Each of these functions of control, improvement and communication is essential for an organization to adapt and develop. However, precisely because evaluative procedures do fulfil these essential functions, they constitute a challenge and create a sense of threat; and, with this, there is the probability of resistance.

As other chapters in this book have repeatedly illustrated, it is never easy for an organization to 'think'. A variety of emotional pressures and anxieties give rise to taken-for-granted routines of the service and mitigate against changing these. It is the return of these feared elements in the work which constitutes the greatest threat embodied in evaluation, as in all organizational interventions. The element of appraisal increases this sense of threat. Any system of evaluation can all too easily come to feel like an accusation of inadequacy: it then comes to represent a critical 'parental' voice, both for the service staff as a whole, and for the individual workers involved.

Defences will inevitably be mobilized against the feelings of exposure and guilt evoked by the introduction of this kind of intervention. These constitute real dangers to the constructive functioning of evaluative systems. Routinized quantitative systems can be used to reduce the need to think about the service to a question of method and technique, an off-the-peg solution to the requirement to undertake evaluation. Indeed, the current fashion for quality assurance can be viewed with some suspicion as an inappropriate attempt to objectify difficult choices about values and priorities, and to dispense with inevitable conflicts and uncertainties by hiding behind the appearance of scientific method. Even when thought and work are put into creating an approach appropriate to an individual service, there is a danger that it will quickly become a matter of routine, rather than a fresh challenge to re-think what has been taken for granted.

A further resistance occurs when the responsibility for appraisal of the service is seen primarily as an exercise in public relations, something that must be seen to be done, rather than being in the interests of the service. Cynicism is increased when external managers also feel threatened and use the evaluation system as a route to exercising inappropriate and un-negotiated controls over services. Thus, there is potential for systems of evaluation to become yet another part of an organization's structured defences against the demands and anxieties of its primary task.

The essence of evaluation, therefore, is the process of learning from experience, a 'process through which institutions think about themselves and their clients' (Clifford *et al.* 1989). This emphasis on reflection, learning and thought stresses that the problem is not so much obtaining information to assess the current service, but getting that information into the 'mind' of the organization in such a way as to enable learning and change.

SUPPORTING COMMITMENT TO LEARNING

Through recognizing these resistances, it becomes possible to think about how to provide the organizational conditions which will enable evaluation to be translated into thoughtful action. We are always faced with a situation in which there is a wish to progress, to improve the service and do better, a desire to learn, but one which is masked by anxiety, the feared consequences of facing up to the realities with which the service deals. It is this desire for growth and the commitment to do good work which must be supported if evaluation is to be effective. Some of the ways in which this might be done were implicit in the approach which was taken at Southside.

Most fundamental is finding a way in which elements of both internal and external review can be incorporated in an evaluation. Thus, an attempt was made to engage staff actively by being honest and direct about what was being asked of them, but then to proceed by addressing their concerns, in this case about lack of communication with management and their sense of being over-burdened. These feelings of stress and confusion were heard sympathetically in the first two meetings and related to the evaluation by linking them to the uncertainty and lack of boundaries around the centre's aims. It was essential at this point to contain the staff's anxiety and their frustration by modelling an attitude of non-judgemental curiosity, and so endeavouring to create a spirit of enquiry. It thus became possible to take the next step of involving staff in the process of gathering data to address the questions about the aims of the service and so perhaps to present their case, as they continued to see it. The evaluation ceased to be something entirely imposed from the outside. At the same time, staff came to accept my role as a sympathetic outsider who could validate their ideas, but might also be able to provide enough challenge to help them confront issues which they found difficult or confusing.

It is easy to become overwhelmed when faced with evaluating complex systems, particularly for staff who are embedded in them. In addition to having outside help, a structure to guide the choice of what to focus on will often be experienced as containing. In the case of Southside, this element was provided by schedules of questionnaires which directed attention to specific problem areas in this type of service (see Leiper *et al.* 1992). It is also essential to be selective and realistic about which aspects of a service to review at any one time, and to resist the wish to produce a 'perfect' service.

The support and involvement of external management is crucial for quality assurance. Managers must legitimize the process, and they may have to consider the provision of new resources in order to correct identified problems. There is a very real danger of evaluation being 'owned' by staff in a way that attempts to pretend that they have complete autonomy and in order to deny management's authority. Equally, a manager may be inclined to attempt to impose systems and so lose staff commitment to implementing any necessary changes. These splitting processes are defences endemic to health and welfare services and must be tackled at the outset, as they were at Southside by initiating a staff group meeting with a previously remote area manager. The aim is to achieve joint ownership and to involve both managers and staff in ways that pay more than lip-service to the process. Quality assurance embodies a process which directly addresses issues which concern both parties. It can thus act as an opportunity to bridge the gap between them and to establish dialogue.

Another split that is commonplace within mental health services is that between staff and the clients receiving services. Stereotypical and denigratory perceptions of service users can easily go unquestioned, and the perceived needs of the people for whom the service is ostensibly provided are then progressively discounted. Consequently, it is helpful to set up structures which create some dialogue with the users of the service. In the case of Southside, the occupational therapist was prepared to take this on; this ran the danger of exacerbating a division between the medical and nursing staff and the others, but provided a starting-point. Again, it is important to try to find ways which make consumer feedback a working reality and avoid a tokenistic approach which can only engender cynicism in the long term.

The guiding principle should be that information is designed to promote action. It should be diagnostic of significant problems, suggesting their causes and pointing to strategies for change. It is for this reason that it is helpful to include information about important service processes, such as team functioning, which are known to influence indirectly the quality of care and to be crucial to implementing change. At the same time, the information should make sense to those who are going to act on it, be communicated in a form which is vivid and have direct implications for the lives of staff and users. Above all, it must indicate problems on which action by those involved can have some impact. Of course, such impact may nevertheless continue to be resisted by some in the service who see it as a threat to their interests or their security.

Southside concluded their review by agreeing that the team co-ordinator and I would co-operate to write a report on the findings, with suggestions for further action which would be discussed first by the staff and then with the area manager. Recommendations included agreeing with managers a statement of priorities for the work which would specify the proportion of staff time devoted to different types of work with the main client groups. This involved a shift in current work priorities towards the more seriously disturbed, and required the development of new information systems to monitor it. As part of this, it was agreed to undertake more joint teamwork with these difficult clients, and to support this by restructuring the team meetings to have separate times for business and clinical issues. A monthly peer review of work with the most difficult clients was initiated. It was proposed that a regular 'users' forum' for the long-term clients be formed to consult them; and information sheets about treatment obtained or written (as they had requested). The manager agreed to support these changes and represent them to local general practitioners, to fund a computerized information system, and to engage a consultant to work with the staff on the issues of relationships and accountability within the team which remained unresolved.

The team had to face some very painful choices in which some staff would have to give up valued areas of their work or change familiar individual working practices. The final meetings were hesitant and involved going over some old ground. Members were helped by their commitment to the relationship with me as their consultant, and by the care taken in the report and the meetings to acknowledge the existing level of skill and commitment shown by staff to their work.

The need for facilitation of the process of evaluation is too often overlooked, particularly at the point where staff as a group are confronting the problems raised by its results. It is at the point of seeking to take corrective action that group processes are likely once again to inhibit change. It is crucial to anticipate such difficulties and to realize that service evaluation, like any other organizational intervention, will not produce major change quickly. It is a continuous, long-term project which requires persistence above all. At Southside, a review date of the action plan was set with me for eight months later, with the intention of going on to review further areas of the service. Quality assurance should not be a once-and-for-all exercise, but a rolling process. The aim should be for appropriate self-evaluation to become part of the culture of the organization.

CONCLUSION: DEVELOPING A SPIRIT OF ENQUIRY

There are a variety of approaches and systems which have been developed for evaluation in recent years. However, methods and materials are of less consequence than the frame of mind in which evaluation is undertaken. It requires the development of a spirit of enquiry and conditions which provide the security necessary for learning from experience, especially containing anxiety, both organizational and personal. The tradition of institutional consultancy described in this book has always striven to provide precisely these elements, by placing a respected outsider in a distinctive position in relation to the system.

While remaining external, he or she seeks a collaborative relationship with staff of the service in order to promote their self-reflection, encourage innovative action and develop the organization's capacity to think and to sustain change.

In moving towards healthier organizations, we endeavour consciously to adopt an orientation towards our real work-tasks and to review working practices and relationships which are familiar and taken for granted. This involves the capacity to take risks and to be open, to have challenged what provides us with security and a familiar sense of coherent identity. It is never likely to be easy. Evaluation can and should be a model of continual learning; evaluators and consultants themselves need to be constantly self-questioning if their work is to develop and to renew itself. Evaluation always challenges our fantasies of omnipotence. It is a hard lesson to learn and a continuing struggle.

Afterword

Anton Obholzer

The theme of this book has been that working in the human services inevitably arouses anxiety, pain and confusion. As a result, institutions, working practices and staff relationships are unconsciously structured so as to defend against anxiety. Furthermore, these defensive structures come to be seen as the optimal way of performing the task and, as such, not to be questioned. We have also seen how institutional defences within helping organizations often exacerbate rather than reduce the stress of working with people in distress. The need for change is therefore self-evident and generally acknowledged. Yet what is also evident is that useful and meaningful change is extremely hard to bring about. Why is this so, and what can we do about it?

ANXIETY AND RESISTANCE TO CHANGE

Human beings are notoriously resistant to change, even when the change appears to be relatively minor or those concerned have ostensibly agreed to it. Managing change inevitably requires managing the anxieties and resistance arising from the change process. It is therefore important to understand the nature of the anxieties that are stirred up, as well as those inherent in the regular work of the organization.

There are, as we see it, three layers of anxiety that need to be understood before they are addressed: primitive anxieties, anxieties arising out of the nature of the work, and personal anxieties.

By primitive anxiety, we refer to the ever-present, all-pervasive anxiety that besets the whole of humankind. A great many authors, philosophers, anthropologists and psychologists have written about it. It is the terror of the baby afraid of being left alone in the dark, of the child who hides under the covers from 'the monsters under the bed', of the person who wears an amulet as protection against the 'evil eye'. Just as Australian Aborigines and Kalahari Bushmen band together and create rituals to hold them together, to provide them with a 'social skin', so do we. And now that many of the social institutions of the past no longer serve to protect us, we imbue existing everyday institutions with protective/defensive functions. They protect us from personal and social breakdown, give us a sense of belonging, save us from feeling lost and alone. Anything that threatens to sever us from the band – redundancy, retirement, migration, institutional change – can flood us with this kind of anxiety.

A second category of anxiety arises from the nature of the work. Many of the chapters in this book have described one or another work-setting, and how the particular nature of the work elicits work-specific anxieties. In response to these anxieties, the work comes to be organized not to pursue the primary task, but rather to defend members of the institution from anxiety. This is done unconsciously, and the defensive function of the structure of the institution usually goes unrecognized. Furthermore, this work-generated anxiety often resonates both with the primitive anxiety discussed above, and with personal anxiety, the anxiety we feel when something triggers off elements of past experience, both conscious and unconscious. Thus, for example, the receptionist at an infertility clinic who has recently had an abortion may experience personal anxiety at work related both to conscious concerns about her possible future infertility and to unconscious fantasies about retribution.

Our need to have 'containing' institutions, that is, for our institutions to protect us from being overwhelmed by these different layers of anxiety, is often at odds with and deflects from the institution's primary task and the changes required to pursue it. The failure to recognize the anxiety-containing function of institutions means that even the best-intentioned organizational changes often create more problems than they solve, since they lead to the dismantling of structures which were erected in the first instance to defend against anxiety. Change, then, not only meets with resistance, but can produce staff illness, breakdown and burn-out. Yet without change, the institution may become increasingly off-task and out of touch with the environment it is supposed to serve, putting its very existence at risk. So, on the face of it, it would appear that attempts at managing change make for a 'damned if we do, damned if we don't' situation.

CONTAINING ANXIETY

We believe that there is an alternative approach. It is not so much a matter of whether to dismantle defensive structures or not, but how, and what to put in their place, and what can be done to reduce the level of disturbance inevitably aroused by change.

In the first place, it is vital for there to be clarity and ongoing discussion about the primary task of the organization, taking into account changes in the environment. In considering any proposed changes, two questions need to be addressed. Will the changes serve the primary task or not? And is the resistance to change part of an attempt to get the organization away from its primary task, or is it a movement to safeguard the task against changes which are actually themselves anti-task?

Second, authority structures need to be clear. Who decides the primary task, and by what authority? It is not uncommon for the staff and management of an institution to have a different view of what their primary task is from the view held by, say, 'headquarters' or 'region' or government ministries. For example, a dispute between the Minister of Education, on one side, and teachers, on the other, can be seen as a dispute about who has authority to determine the primary task and how it is to be achieved.

It is essential that clear and open communication between all the sectors involved be maintained: there needs to be a forum where these issues can be debated. This applies equally to in-house meetings of a small organization, and to large-scale inter-institutional meetings. It is sometimes argued that, under the daily pressure of ensuring that tasks are accomplished, debate is time-wasting. However, it is my view that managing without a forum for debate, that is, without consultation, can only contribute to greater institutional malaise.

In addition to a forum for debate, we need to have work-related staff support systems to contain the anxieties arising from the work itself, as well as out of the process of change. In the human services, the more on-task the working structure of the organization, the more the staff are likely to feel they are working on a bed of nails, the 'nails' being the work-related pain that punctures their sensibilities. Systems which encourage open discussion of work-related feelings and problems, in a climate which regards 'problems' as a normal aspect of the work rather than as evidence of personal or group pathology, encourage individual and organizational learning, and thus foster growth and development. This kind of climate can enhance the support function of existing systems, such as supervision, client reviews and the like. In many cases, however, it is useful, even necessary, to have also specially designated and regularly time-tabled meetings devoted to reviewing and reflecting on work tensions and how these are affecting the staff and their practice. The presence of an external consultant to such meetings can help protect the reflective space from being eroded either by other work pressures, or by group defences.

These kinds of reflective, exploratory discussions can alert managers and members when the toll taken by the work is too high and something needs to be done. For example, it may be that certain roles and placements should be time-limited. In nuclear medicine departments, it is recognized that staff can safely receive only a certain amount of radiation over a limited period. Once the radiation-counter on their lapel badge goes over that amount, they have to take time off. Similarly, it needs to be recognized that in certain care-settings associated with particularly high levels of stress, staff will and should move on. Instead, losing staff is usually regarded as a catastrophe, or as evidence of failure.

In addition, personal support systems should be available, for example, through a confidential staff counselling service. However, care needs to be taken not to attribute to 'personal problems' issues which properly belong to the whole group. The individual may then be 'holding' or expressing a difficulty on behalf of others, and supporting them 'personally' may become a way to prevent the necessary organizational reflection and change taking place.

Finally, managers also need support. The absence of systems of managerial support can make for loneliness in the managers. The collusive hand of companionship is likely then to be doubly welcome, and managers' capacity to fulfil their containment and leadership functions will be undermined.

Throughout this book we have tried to demonstrate the potential usefulness, not only for managers but also for staff struggling to manage themselves in their work-roles, of having an awareness and understanding of unconscious institutional processes. This can be invaluable in helping to make sense of how one is perceived and treated, how one feels at work, and how this has to do not only with one's role, but also with projections and with re-enactments of one's own past. It can also help to reduce scapegoating of groups and individuals, and rigid adherence to traditional work practices and assumptions. Thus, it can contribute much to maintaining a healthier working climate in the human services, where anxiety, pain and confusion are inevitable. Since changes in these services are also inevitable, and on an ever greater scale, increasingly skilful and sensitive management is needed.

SOME COMMENTS ON THE CONSULTANCY ROLE

It is generally recognized that if you want to refurbish a building, you need expert advice to locate load-bearing walls, the distribution of services and so on. Neglecting this can, in the process of alteration, bring the house down on your head. Yet when it comes to changing organizations, a much more cavalier attitude prevails. If the institutional equivalent of a load-bearing wall comes down, or a short-circuit is produced, it is treated as if it were either inevitable or bad luck.

Consultants to institutions can be regarded as having an analogous role to the architect's, predicting which are the load-bearing structures, and helping to identify what sort of emotional loads these structures are carrying. It is essential that the implied question 'What are you going to do about this, and who, in what role, is going to take the lead in this work venture?' is an integral part of the exploration of institutional stresses and defences. It is important that the solution of the problem arises from a collaboration between the consultant and those within the organization, taking into account their management style and language. This way, the risk of undermining the existing management is reduced, as is the risk of bringing about temporary improvement, followed by collapse on withdrawal of the consultant.

The institution, therefore, is best served by a form of consultancy which does not have a preconceived idea of what the structure of the organization should be on completion of the intervention, while giving the consultant opportunities to communicate ideas as they arise. The outcome, instead, should be determined by a public process of striving towards understanding. With this comes awareness that the task of monitoring and reviewing is never complete and needs to be supported in an ongoing way. The consultant who offers a psychodynamic understanding of institutional process also brings a state of mind and a system of values that listens to people, encourages thought and takes anxieties and resistance into account. At the end of the consultation, the organization will, one hopes, have taken this stance into its culture: a new awareness of the potential risks to the work and the workers as a result of the stresses inherent in the organization's task, as well as of the costs of neglecting these, together with greater clarity about how to proceed.

My favourite definition of consultancy is 'licensed stupidity'. The consultant, as a non-member of the institution, is in a position to ask naïve questions about the reasons for the existence of a variety of structures and policies. These often seem perfectly obvious to members, but may actually be the outcrops of the reef of institutional defensive processes. This book was written as an invitation to readers to stand back and look at their own situation with fresh eyes, to question the 'obvious', and to make time and space for reflecting on their observations and experiences.

Appendix

The following is a list of some of the organizations in the United Kingdom and abroad which offer group relations training and/or consultancy influenced by the conceptual frameworks described in this book. Inevitably, it is incomplete and includes only those organizations of whose activities we are aware. It is not intended in any way either to 'accredit' those listed, or indeed to discredit those omitted.

Australia
Australian Institute of Social Analysis, 21 Drummond Place, Caroton, Victoria 3053.

Belgium
International Forum for Social Innovation/Forum International de L'Innovation Sociale, 56 rue du Lieutenant Pirard, 45 44 Dalhem.

Denmark
Proces Aps, Valnuddegarden 5, DK 26 60 Albertslund.

Finland
Finnish Society for Organizational Dynamics, Koskitie 45B7, 90500 Oulu.

France
International Foundation for Social Innovation, 16 avenue de Friedland, 75008 Paris.
L'Institut pour la Conduite du Changement Social, Siège Social, PB 1352, 49013 Angers Cedex 01.

Germany
Mundo, Wotanstrasse 10, D-5600 Wuppertal.

India
Indian Institute of Management Calcutta, Post Box 16757, Alipore Post Office, Bengal.
SAKTI, 1357 9th Cross, First Place, JPNagar, Bengalore 560078.

Israel
Innovation and Change in Society, The Israel Management Center, POB 3033, Tel Aviv 61 330.
Israel Association for the Study of Group and Organizational Processes, Department of Social Work, Hebrew University of Jerusalem, Mount Scopus, Jerusalem 91 905.

Italy
ISMO, Piazza San Ambrogio, 20123 Milan.
Studio Analisi PsicoSociologica, Corso Vercelli 58, 20145 Milan.

Mexico
Instituto Mexicano de Relaciónes Grupalales y Organicacionales, Hacienda de Santa Ana y Lobos 26, Bosque de Echegarai, Estado de Mexico 53310.

Netherlands
ISI, Westerboutstraat 48, N.2012 Haarlem.

Norway
Norstig, ULA-NTH 7034, Trondheim.

South Africa
Institute for the Study of Leadership and Authority in South Africa, PO Box 131 334, Bryanston 2021.

Sweden
Agslo (Arbetsgruppen for Studium av Ledarskap och Organisation), Box 60, 126, 21 Hagersten, Stockholm.

United Kingdom
The Grubb Institute, Cloudesley Street, London N1 OHU.
OPUS Consultancy Services, 10 Golders Rise, London NW4 2HR.
Scottish Institute of Human Relations, 21 Elmbank Street, Glasgow G2 4PE.
The Tavistock Clinic Foundation, Tavistock Centre, 120 Belsize Lane, London NW3 5BA.
The Tavistock Institute of Human Relations, Tavistock Centre, 120 Belsize Lane, London NW3 5BA.

United States of America
A.K. Rice Institute, PO Box 39102, Washington, DC 20016.
A.K. Rice Institute, Center for the Study of Groups and Social Systems (Boston Center), 2 Townly Road, Watertown, MA 02172.
A.K. Rice Institute, Central States Center, Rte 2, 7371 HiWay WW, Columbia, MO 65201.
A.K. Rice Institute, Chicago Center for the Study of Groups and Organizations, 2221 Harrison Street, Evanston, IL 60201.
A.K. Rice Institute, GREX (California), PO Box 229, Tiburon, CA 94920.
A.K. Rice Institute, Midwest Group Relations Center, 400 Dyer Hall, University of Cincinnati, Cincinnati, OH 45221-0376.
A.K. Rice Institute, New York Center, PO Box 20416, Park West Station, New York, NY 10025-1513.
A.K. Rice Institute, Texas Center, 3206 Aberdeen Way, Houston, TX 77057.
A.K. Rice Institute, Washington–Baltimore Center, 1610 New Hampshire Avenue, NW, Washington, DC 20009.
Center for Education in Groups and Organizations US, 6505 Wiscasset Road, Bethesda, MD 20816.
William Alanson White Institute, 20 West 74th Street, New York, NY 10023.

Bibliography

Adams, A. and Crawford, N. (1992) *Bullying at Work*, London: Virago.

Anderson, R. (ed.) (1992) *Clinical Lectures on Klein and Bion*, London/New York: Tavistock/Routledge.

Armstrong, D. (1991) 'Thoughts bounded and thoughts free', paper to Department of Psychotherapy, Cambridge.

—— (1992) 'Names, thoughts and lies: the relevance of Bion's later writings for understanding experiences in groups', *Free Associations*, 3.26: 261–82.

Bain, A. (1982) *The Baric Experiment: the Design of Jobs Organization for the Expression and Growth of Human Capacity*, Tavistock Institute of Human Relations Occasional Paper no. 4.

Balint, M. (1964) *The Doctor, His Patient and the Illness*, London: Pitman Medical.

Bion, W. (1961) *Experiences in Groups*, New York: Basic Books (see 'Selections from: Experiences in Groups', in A. D. Colman and W.H. Bexton (eds) *Group Relations Reader 1*, A. K. Rice Institute Series [Washington, DC], 1975).

—— (1962) 'Learning from experience', *International Journal of Psychoanalysis*, 43: 306–10.

—— (1967) 'Attacks on linking', in *Second Thoughts: Selected Papers on Psychoanalysis*, London: Heinemann Medical (reprinted London: Maresfield Reprints, 1984).

—— (1977) *Seven Servants*, New York: Jason Aronson.

—— (1980) *Bion in New York and São Paulo*, (ed.) F. Bion, Perthshire: Clunie Press.

—— (1984) *Elements of Psychoanalysis*, London: Heinemann.

Bottoms, A.E. and McWilliams, W. (1979) 'A non-treatment paradigm for probation practice', *British Journal of Social Work*, 9.2: 159–202.

Bridger, H. (1990) 'Courses and working conferences as transitional learning institutions', in E. Trist and H. Murray (eds) *The Social Engagement of Social Science, Volume 1: The Socio-Psychological Perspective*, London: Free Association Books.

Cardona, F. (1992) 'Crescere, educare, curare contenere; strutture residenziali per adolescenti in grave difficoltà in Inghilterra', in C. Kaneklin and A. Orsenigo (eds) *Il lavoro di comunità*, Rome: La nuova Italia scientifica.

Clifford, P., Leiper, R., Lavender, A. and Pilling, S. (1989) *Assuring Quality in Mental Health Services: the QUARTZ System*, London: RDP/Free Association Books.

Colman, A. D. (1975) 'Irrational aspects of design', in A.D. Colman and W.H. Bexton (eds) *Group Relations Reader 1*, A. K. Rice Institute Series [Washington, DC].

Daniell, D. (1985) 'Love and work: complementary aspects of personal identity', *International Journal of Social Economics*, 12.2: 48–55.

Dearnley, B. (1985) 'A plain man's guide to supervision – or new clothes for the emperor?', *Journal of Social Work Practice*, 2.1, November: 52–65.

Dicks, H. (1970) *Fifty Years of the Tavistock Clinic*, London: Routledge & Kegan Paul.

Fabricius, J. (1991) 'Running on the spot or can nursing really change?', *Psychoanalytic Psychotherapy*, 5.2: 97–108.

Fletcher, A. (1983) 'Working in a neonatal intensive care unit', *Journal of Child Psychotherapy*, 9.1: 47–55.

Freud, S. (1917) 'Mourning and melancholia', in *Collected Papers*, vol. 4, London: Hogarth Press, 1925.

—— (1921) *Group Psychology and the Analysis of the Ego*, Penguin Freud Library, vol. 12, Harmondsworth: Penguin Books, 1984.

—— (1924) 'Recommendations to physicians practising psychoanalysis', in *Standard Edition*, vol. 12, London: Hogarth Press, 1958.

Griffiths, R. (1988) *Community Care: Agenda for Action*, London: HMSO.

Grubb Institute (1991) 'Professional management', notes prepared by the Grubb Institute on concepts relating to professional management.

Gustafson, J. P. (1976) 'The pseudomutual small group or institution', *Human Relations*, 29: 989–97.

Hinshelwood, R.D. (1989) *A Dictionary of Kleinian Thought*, London: Free Association Books.

Hirschorn, L. (1988) *The Workplace within: Psychodynamics of Organizational Life*, Cambridge, MA: MIT Press.

Home Office (1990) *Supervision and Punishment in the Community*, London: HMSO.

Hornby, S. (1983) 'Collaboration in social work: a major practice issue', *Journal of Social Work Practice*, 1.1:35–55.

Jaques, E. (1948) 'Interpretive group discussion as a method of facilitating social change', *Human Relations*, 1: 533–49.

—— (1951) *The Changing Culture of a Factory*, London: Tavistock Publications (see 'Working-through industrial conflict: the service department at the Glacier Metal Company', in E. Trist and H. Murray (eds) *The Social Engagement of Social Science, Volume 1: The Socio-Psychological Perspective*, London: Free Association Books, 1990).

—— (1953) 'On the dynamics of social structure: a contribution to the psychoanalytical study of social phenomena deriving from the views of Melanie Klein', in E. Trist and H. Murray (eds) *The Social Engagement of Social Science, Volume 1: The Socio-Psychological Perspective*, London: Free Association Books, 1990.

—— (1965) 'Death and the mid-life crisis', in E. B. Spillius (ed.) *Melanie Klein Today, Volume 2: Mainly Practice*, London: Routledge, 1990.

Jervis, M. (1989) 'Radical reforms and golden opportunities', *Social Work Today*, March: 12–13.

Klein, L. and Eason, K. (1991) *Putting Social Science to Work*, Cambridge: Cambridge University Press.

Klein, M. (1959) 'Our adult world and its roots in infancy', in A. D. Colman and M. H. Geller (eds) *Group Relations Reader 2*, A. K. Rice Institute Series [Washington, DC], 1985.

Lawrence, G. (1977) 'Management development . . . some ideals, images and realities', in A. D. Colman and M. H. Geller (eds) *Group Relations Reader 2*, A. K. Rice Institute Series [Washington, DC], 1985.

Leiper, R., Lavender, A., Pilling, S. and Clifford, P. (1992) *Structures for the Quality Review of Mental Health Settings: the QUARTZ Schedules*, Brighton: Pavilion.

Lewin, K. (1947) 'Frontiers in group dynamics, Parts I and II', *Human Relations*, 1: 5–41; 2: 143–53.

Main, T. (1968) 'The ailment', in E. Barnes (ed.) *Psychosocial Nursing: Studies from the Cassel Hospital*, London: Tavistock Publications.

Mattinson, J. (1975) *The Reflection Process in Casework Supervision*, London: Institute of Marital Studies.

—— (1981) 'The deadly equal triangle', in Smith College School of Social Work/Group for the Advancement of Psychotherapy in Social Work, *Change and Renewal in Psychodynamic Social Work: British and American Developments in Practice and Education for Services to Families and Children*, Northampton, MA/London: SCSSW and GAPS.

Mattinson, J. and Sinclair, I. (1979) *Mate and Stalemate: Working with Marital Problems in a Local Authority Social Services Department*, Oxford: Blackwell.

Meltzer, D. (1978) *The Kleinian Development*, Perthshire: Clunie Press.

Menzies, I. E. P. (1960) 'Social systems as a defence against anxiety: an empirical study of the nursing service of a general hospital', in E. Trist and H. Murray (eds) *The Social Engagement of Social Science, Volume 1: The Socio-Psychological Perspective*, London: Free Association Books, 1990.

Menzies Lyth, I. E. P. (1979) 'Staff support systems: task and anti-task in adolescent institutions', in *Containing Anxiety in Institutions: Selected Essays*, London: Free Association Books, 1988.

—— (1983) 'Bion's contribution to thinking about groups', in J. S. Grotstein (ed.) *Do I Dare Disturb the Universe?*, London: Maresfield Library.

—— (1990) 'A psychoanalytical perspective on social institutions', in E. Trist and H. Murray (eds) *The Social Engagement of Social Science, Volume 1: The Socio-Psychological Perspective*, London: Free Association Books.

Millar, D. and Zagier Roberts, V. (1986) 'Elderly patients in "continuing care": a consultation concerning the quality of life', *Group Analysis*, 19: 45–59.

Miller, E. J. (1990a) 'Experiential learning in groups I: the development of the Leicester model', in E. Trist and H. Murray (eds) *The Social Engagement of Social Science, Volume 1: The Socio-Psychological Perspective*, London: Free Association Books.

—— (1990b) 'Experiential learning in groups II: recent developments in dissemination and application', in E. Trist and H. Murray (eds) *The Social Engagement of Social Science, Volume 1: The Socio-Psychological Perspective*, London: Free Association Books.

Miller, E. J. and Gwynne, G. (1972) *A Life Apart*, London: Tavistock Publications.

Miller, E. J. and Rice, A. K. (1967) *Systems of Organization: The Control of Task and Sentient Boundaries*, London: Tavistock Publications (see 'Selections from: Systems of Organization', in A. D. Colman and W. H. Bexton (eds) *Group Relations Reader 1*, A. K. Rice Institute Series [Washington, DC], 1975; see also 'Task and sentient systems and their boundary controls', in E. Trist and H. Murray (eds) *The Social Engagement of Social Science, Volume 1: The Socio-Psychological Perspective*, London: Free Association Books, 1990).

Nightingale, F. (1860) *Notes on Nursing: What It Is and What It Is Not*, New York: Dover Publications, 1969.

Obholzer, A. (1987) 'Institutional dynamics and resistance to change', *Psychoanalytic Psychotherapy*, 2.3: 201–5.

Reed, B. D. and Palmer B. W. M. (1972) *An Introduction to Organizational Behaviour*, London: Grubb Institute.

Rice, A. K. (1963) *The Enterprise and Its Environment*, London: Tavistock Publications.

—— (1965) *Learning for Leadership*, London: Tavistock Publications (see 'Selections from: Learning for Leadership', in A. D. Colman and W. H. Bexton (eds) *Group Relations Reader 1*, A. K. Rice Institute Series [Washington, DC], 1975).

—— (1969) 'Individual, group and inter-group processes', in E. Trist and H. Murray (eds) *The Social Engagement of Social Science, Volume 1: The Socio-Psychological Perspective*, London: Free Association Books, 1990.

Santayana, G. (1905) *The Life of Reason, Volume 1* (republished by Dover 1980).

Segal, H. (1957) 'Notes on symbol formation', in *The Work of Hanna Segal: a Kleinian Approach to Clinical Practice*, London: Free Association Books/Maresfield Library, 1986.

—— (1977)'Psychoanalysis and freedom of thought', in *The Work of Hanna Segal: a Kleinian Approach to Clinical Practice*, London: Free Association Books/Maresfield Library 1986.

—— (1979) *Klein*, London: Fontana.

—— (1986) 'Manic reparation', in *The Work of Hanna Segal: a Kleinian Approach to Clinical Practice*, London: Free Association Books/Maresfield Library.

Skynner, R. (1989) *Institutes and How to Survive Them*, London: Methuen.

Symington, N. (1986) *The Analytic Experience: Lectures from the Tavistock*, London: Free Association Books.

Trist, E., Higgin, G., Murray, H. and Pollock, A. (1963) *Organizational Choice*, London: Tavistock Publications (see shortened version 'The assumption of ordinariness as a denial mechanism: innovation and conflict in a coal mine', in E. Trist and H. Murray (eds) *The Social Engagement of Social Science, Volume 1: The Socio-Psychological Perspective*, London: Free Association Books, 1990).

Turquet, P. (1974) 'Leadership: the individual and the group', in A. D. Colman and M. H. Geller (eds) *Group Relations Reader 2*, A. K. Rice Institute Series [Washington, DC], 1985.

Wells, L. (1985) 'The group-as-a-whole perspective and its theoretical roots', in A. D. Colman and M. H. Geller (eds) *Group Relations Reader 2*, A. K. Rice Institute Series [Washington, DC].

Winnicott, D. W. (1947) 'Hate in the countertransference', in *Collected Papers: through Paediatrics to Psycho-analysis*, London: Hogarth Press and the Institute of Psycho-analysis, 1958.

—— (1971) *Playing and Reality*, London: Tavistock Publications (reprinted Harmondsworth: Penguin Books 1980).

Wollheim, R. (1971) *Freud*, London: Fontana.

Woodham-Smith, C. (1950) *Florence Nightingale, 1820–1910*, London: Constable.

Woodhouse, D. and Pengelly, P. (1991) *Anxiety and the Dynamics of Collaboration*, Aberdeen: Aberdeen University Press.

Suggestions for further reading

Anderson, R. (ed.) (1992) *Clinical Lectures on Klein and Bion*, London/New York: Tavistock/Routledge.

Colman, A. D. and Bexton, W. H. (eds) (1975) *Group Relations Reader 1*, A. K. Rice Institute Series [Washington, DC].

Colman, A. D. and Geller, M. H. (eds) (1985) *Group Relations Reader 2*, A. K. Rice Institute Series [Washington, DC].

De Board, R. (1978) *The Psychoanalysis of Organizations*, London: Tavistock Publications.

Hinshelwood, R. D. (1987) *What Happens in Groups: Psychoanalysis, the Individual and the Community*, London: Free Association Books.

Menzies Lyth, I. E. P. (1988) *Containing Anxiety in Institutions: Selected Essays*, London: Free Association Books.

—— (1989) *The Dynamics of the Social: Selected Essays*, London: Free Association Books.

Miller, E. J. (1993) *From Dependency to Autonomy: Studies in Organization and Change*, London: Free Association Books.

Segal, J. (1985) *Phantasy in Everyday Life: a Psychoanalytical Approach to Understanding Ourselves*, Harmondsworth: Penguin Books.

—— (1992) *Melanie Klein*, London: Sage Publications.

Trist, E. and Murray, H. (eds) (1990) *The Social Engagement of Social Science, Volume 1: The Socio-Psychological Perspective*, London: Free Association Books.

Index

Index entries in italics refer to case illustrations.